Thomas Chesmer Weston

Reminiscences among the Rocks

In Connection with the Geological Survey of Canada

Thomas Chesmer Weston

Reminiscences among the Rocks
In Connection with the Geological Survey of Canada

ISBN/EAN: 9783337205881

Printed in Europe, USA, Canada, Australia, Japan

Cover: Foto ©berggeist007 / pixelio.de

More available books at **www.hansebooks.com**

REMINISCENCES AMONG THE ROCKS

IN CONNECTION WITH THE GEOLOGICAL SURVEY OF CANADA

BY

THOS. CHESMER WESTON, F.G.S.A.

TORONTO
PRINTED FOR THE AUTHOR BY
WARWICK BRO'S & RUTTER
1899

PREFACE

AT the request of my children and several of my friends, I have endeavoured to record some of the incidents connected with my explorations while in the service of the Government Geological Survey of the Dominion of Canada. I have also set down a few details of the geological formations in which I have worked. I have prefixed to these writings some autobiographical particulars which may not be without interest to the reader. It is a very plain tale, without the slightest striving after any of the literary properties that generally accompany a story.

T. C. W.

237 Daly Avenue, Ottawa.
    May 7th, 1898

CONTENTS

CHAPTER I.

A few autobiographical particulars—Proposal that I should come to Canada—The voyage—First impressions of the land of the maple—A matrimonial joke—My introduction to the rocks 1

CHAPTER II.

Eozoon—"The Dawn of Life"—A remarkable geological controversy—Discovery of tubuli structure—Unlikely that the extensive Laurentian system would only contain one kind of organic remains—That the Huronian system has not yielded any fossils 20

CHAPTER III.

A visit to the United States—Journey to the Island of Anticosti—My assistant proves forgetful and unfortunate—Description of the island—Its geological formation—Some queer inhabitants—I sight a couple of bears—Further misfortunes of Jackson—Departure for Quebec............... 28

CONTENTS.

CHAPTER IV.

The South Petite Nation River—The Trenton formation—The calciferous rocks of Ontario—We lose our way in the bush—The Chazy formation—A Guelph freethinker—An adventure with a bull—Visit to Point Levis—An odd mistake at our boarding-house—The Eozoon controversy 46

CHAPTER V.

The field season of 1869—Journey with Mr. Richardson—Lost in the bush—The primordial rocks in New Brunswick—Arisaig, Nova Scotia, and its coast rocks—Retirement of Sir William Logan—Field season of 1870—The rocks of Quebec—Another visit to the United States—Visit to Stanfold, Que.—Field work of 1871-2—The coast of Labrador 59

CHAPTER VI.

Lord Dufferin visits the Survey—I assist Sir William Logan in the Eastern Townships—Another visit to Arisaig—Some of the pioneer photographs of the survey—A storm off Cape Gaspé—A visit to Newfoundland — Lake Temiscouata — Work at Hespeler, Ont.—The conglomerates of Bic—Death of Sir William Logan—The Philadelphia exhibition—The Eastern Townships again—Distribution of rocks and minerals—Ethnological branch of the Survey........ 79

CHAPTER VII.

The south shore of the St. Lawrence—Some good fishing—Camping by the river—Ascent of the Shickshocks—A poaching expedition: we catch a fine salmon—The lonely Mount Albert—Photographs and fossils—Coast rocks between Digby and Yarmouth, N.S.—" The Ovens "—Fishing at Spindle Cove—Lost in the fog 107

CONTENTS.

CHAPTER VIII.

Baie des Chaleurs and adjacent districts—A fossil fish and a pious critic—A list of tithes—Removal of the headquarters of the Survey from Montreal to Ottawa—The Joggins Coast, N.S.—Campbellton and the Restigouche River—A trip to Sault Ste. Marie—The Cypress Hill and the North-West—A three months geological exploring tour 127

CHAPTER IX.

To South Joggins, N.S.—Palæontology of the carboniferous rocks—Journey to Manitoba and the North-West—Maple Creek and the Cypress Hills—Vertebrate remains—A herd of wild cattle—Belly River and the prairies—Some Indians—Selkirk and its attractions—A pelican—The Red River—Winnipeg and its miscellaneous stores.... 176

CHAPTER X.

Excavations at Quebec—Eozoon again—A collection of mammalian remains—Work on the Quebec group—The black limestones of Quebec—Supposed coal deposits — Evading the Scott Act — With Mr. Fletcher to Cape Breton—Arisaig, N.S.—The coal mines of Sydney 208

CHAPTER XI.

Explorations on the south shore of the St. Lawrence—Fishing for cod—A pleasant stay at Mr. Richardson's—A journey to Grimsby—A trip to the North-West plains—Maple Creek—Swift Current—Big Jaw coulee—A prize pig—The bad lands—Lethbridge—We launch a boat on the Red Deer River but afterwards give up our plan—The Joggins coast once more 226

CONTENTS.

CHAPTER XII.

Supposed fossil trees — A visit of investigation to Kingston—I start for the Red Deer River—Mr. McKenzie joins me at Calgary—Extensive coal seams—Fine dinosaur remains—Canada has done comparatively little towards restoring these denizens of a silent world—Dangerous quicksands—We celebrate Dominion Day—A long river trip—Another trip to the Cypress Hills—Unpleasant experiences—On the borders of Lake Superior.... 247

CHAPTER XIII.

The "Quebec Group" controversy—An interruption by illness—I afford amusement to the natives—The Montmorency rocks—The gold-bearing rocks of Nova Scotia—Improvements in the museum of the Survey—A clerical geologist—The pilgrims to Bonne Sainte Anne—The geology of the Sainte Anne district—Surrounded by the tide 286

CHAPTER XIV.

Further geologizing at Quebec City—Examining the rocks precipitated by the land-slide—A victim of poison-ivy—A search for a meteorite on the Ottawa River—The Laurentian marble—An artist's sketch is useful—The Joggins coast, N.S., again—Conclusion and thanks to my friends of the Survey. 311

LIST OF ILLUSTRATIONS

T. C. Weston		Frontispiece
Horace S. Smith	facing page	4
Sir W. E. Logan	" "	14
T. Sterry Hunt	" "	18
Eozoon Canadense Canal System	" "	24
Eozoon Canadense Laminated Structure	" "	24
James Richardson	" "	34
E. Billings	" "	70
Alex. Murray	" "	90
Shipping Buffalo Bones	" "	144
Section of Miocene Tertiary Rocks	" "	182
Concretions at Irwin Coulee, N.W.T.	" "	192
Tree-like Concretions at Kingston, Ont	" "	248
Huronian Concretions	" "	292
Concretions, Cape Breton	" "	292

CHAPTER I.

A FEW AUTOBIOGRAPHICAL PARTICULARS—PROPOSAL THAT I SHOULD COME TO CANADA—THE VOYAGE—FIRST IMPRESSIONS OF THE LAND OF THE MAPLE—A MATRIMONIAL JOKE—MY INTRODUCTION TO THE ROCKS.

MY first memories are of Birmingham, England, the "toy shop of Europe," the birth-place of Watt's steam engine, the source of "Brummagem" jewellery and Radicalism. In this town of various manufactures—it was not a city then—I was born in October, 1832. My father was a self-taught natural philosopher in a modest way, who was enthusiastic in learning about science, such as it was in those days. He was particularly fond of optics, and constructed both magic lanterns and microscopes for his own use. His versatility and cleverness in handicraft led him ultimately into the manufacture of fancy ornaments such as women delight in. Part of his business consisted in dealing in and cutting precious stones. He also prepared microscopic sections of stones and minerals. It was this work that earned him the title of "scientific lapidary," and it was in this line of duties that I spent ten years of my early life. Amid the various ups and downs of a busy commercial life my father still pursued his scientific investigations and his optical experiments. One of my earliest remembrances is of finding myself in a large building in which were gathered between one and two thousand school children who had each paid a penny to come and see one of his magic lantern exhibitions—a wonder in those days. The place was in darkness except for the light reflected from a white circle on an immense hanging sheet. Then a picture of the earth flashed on to this space with ships sailing around it altogether disproportionate in size. My father explained to his audience the rotundity of the globe, and concluded his elucidation by "giving out" the first verse of Isaac Watts' hymn :—

    O Spirit of the Lord prepare,
      All the round earth her God to meet,
    Breathe Thou abroad like morning air,
      Till hearts of stone begin to beat.

My father's science did not deprive him of his religious faith. I could linger long over his efforts to combine lecturing and natural philosophy with business, his teaching working men and women and children astronomy, natural history and the like. My education, meanwhile, was not of the most regular kind. But I had a sweet and cultured mother who taught me much. I learned the art of the lapidary, and the time came when we were the most prominent firm in that line in Birmingham.

In January, 1858, my father received from Sir W. E. Logan, Director of the Geological Survey

AMONG THE ROCKS.

of Canada, an offer to come to Montreal as lapidary to the Survey. For various reasons he could not accept this, and I was recommended as competent to fill the position. After some correspondence Sir William decided to engage my services. A sum of money was placed to my credit in one of the Birmingham banks to defray my expenses to Portland, where I was to receive a pass over the Grand Trunk Railway to Montreal. After a very stormy voyage on the Allan liner *Indian*, I landed at Portland on February 15, 1859. I left England with a "stove pipe" hat and a light overcoat, and almost my first experience of a Canadian winter was a frozen ear.

Arriving at Longueuil, opposite Montreal, the scene was so different from anything I had seen —the crowd of sleighs, the jingling of bells, the fur-clad drivers and the hundreds of people who had come to see the passengers of the ship which had been given up for lost, for we were many days behind time—attracted my attention so much that I crossed the St. Lawrence without knowing that I was travelling over a great body of frozen water. The weather was so intensely cold that when I presented a letter of introduction to Mr. Blackwell, director of the Grand Trunk Railway, he advised me to at once get a fur cap and other warm clothing, and directed me to the Geological Survey. I am much indebted to the late director of the Grand Trunk Railway—Mr. Blackwell—for many personal favors.

Arriving at the Survey and Museum, St. Gabriel Street, I found Sir Wm. Logan had gone to Toronto to interview the Government on survey matters. I was however received kindly by Dr. Sterry Hunt, chemist to the survey, and told to rest myself till Sir William arrived, which he did in a few days, and exercised much pains in getting me comfortably settled in the building. I soon arranged my instruments for work, and through the kindness of the Rev. Dr. Wilkes of Zion Church, to whom I brought letters of introduction, was soon at home in my adopted country and thus commenced what lengthened out to thirty-five years of active life in the office, in the museum, and in the field.

On the 9th of June, 1859, (five months after my arrival in Canada) I was married to Matilda Allen, youngest daughter of Smith Allen, of Quebec. We were married by the Rev. Mr. Bancroft, at old Trinity Church. Mr. Horace S. Smith, artist to the Survey, acted as my best man. Horace was always fond of fun and here on this (to me) serious occasion could not help indulging in his " little joke." Just as we were arranged at the altar and the minister was about to commence the marriage service, Smith whispered to me—" I've forgotten the ring." I made up my mind at once that Smith should get out of

Horace S. Smith

AMONG THE ROCKS.

the matter as best he could, and the service went on; but when the time came for the ring Horace quietly took it from his pocket and handed it to the parson. Smith remained a constant friend and I attended him in his last hours of sickness and saw him safely conveyed to his last resting place.

Many interesting incidents occured during my second year in Canada (1860) which have become historical. Among these I will only mention that of the Prince of Wales' visit to Canada, where at Montreal he opened the Victoria Bridge, that wonderful structure which spans our noble St. Lawrence River, and afterwards visited the Geological Survey and took great interest in the rocks and minerals of Canada especially in the seven pound weight of native gold from the Chaudiere valley in the Province of Quebec, which as on many other occasions passed through my hands, from the safe to the cases of the museum.

Through the kindness of Sir William I was introduced to His Royal Highness and accompanied him through the Museum.

From Montreal the Prince of Wales went to Ottawa where on the 1st of September accompanied by the Governor-General, Sir Edmund Head, and a brilliant staff of notables, he laid the foundation stone of the Parliament Buildings.

A short time after my arrival at the Survey,

REMINISCENCES

I found that my duties would consist of developing fossils and reducing the rock in which they were imbedded, which was done by a thin iron disc charged with diamond dust. By this means hundreds of new fossils were placed in the already crowded cases. Microscopic sections of various rocks and fossils were made for study and I was constantly called on to assist Mr. E. Billings, palæontologist to the Survey. This work occupied my whole attention till the early part of 1863 when I commenced my field explorations by accompanying Mr. J. Richardson and Mr. Bell, both members of the geological staff, to Drummondville. Here I was to assist in the examination of the rocks, and collect, if possible, fossils from certain members of what Sir W. E. Logan had termed " The Quebec Group," a series of strata I shall often speak of in these rambling notes.

This was my first experience in field geology, and also my first insight into Canadian country life.

As these pages are to contain beside geological notes, incidents " by the way side," I shall commence here by relating one boarding-house experience. Our home for a few days was situated on the bank of the St. Francis River, Province of Quebec. It is a dear old homestead looking out over the falls of Drummondville, and was kept by an old man and his wife, two thoroughly Scotch bodies who had reared a large family, had seen much of the hard pioneer life of the back woods of Canada and had become somewhat doubtful of one another's veracity as the following narration shows.

Mr. Richardson was a thorough Scotchman, and like many of his class delighted in his glass of toddy before retiring for the night, so before commencing field operations, he provided himself with a stout bottle of Scotch whiskey (his favorite beverage) and having no safe place to keep his bottle, decided to trust it with the old man of the house. The first night, after a hard day's work among the rocks of the river, he called on the old man for his bottle, but on taking his usual "nip" found that the strength of the liquor was not what the label represented it to be. The second evening it was still weaker, and Mr. R. asked the old lady whether her husband was to be trusted with the bottle—" Ah man " she replied " ne'er trust the old man wi' yer whuskey for he will help himself and then fill up wi' water,' 'gi' me the bottle and it will be all richt." So the bottle was given to the old lady for safe keeping ; but alas ! on the following night the whiskey was still weaker, so the old man was then interviewed as to his wife's honesty, " Ah man " he replied " you should na trust the auld woman wi' the whuskey for she helps her sell and then waters the bottle." So concluding they were two of a kind Mr. R. decided to hide the bottle in our bed-room and from that time the strength of the whiskey was preserved.

The rocks of the St. Francis River in the vicinity of Drummondville are chiefly greenish sandstones, forming in places high walls, and showing beautiful examples of stratification and folding; they are underlaid by dark shales and limestones holding a few graptolites and obscure bivalves among which the following genera and species were found:

Dendrograptus simplex, Walcott,
Dicranograptus sextans, Hall,
Cœnograptus gracilis, Hall,
Climacograptus Dicornis, Hall,
Leptobolus insignis, Hall.

These sandstones, limestones and shales are cut by a variety of diorites and amygdaloids, some of which are porphyritic; they form the rocks of the falls, giving a rugged aspect to this portion of the river. It was here among the dark colored shales and limestones of the St. Francis River, I learned from my friend Mr. Richardson—who afterwards during our twenty years of almost constant companionship taught me much of the knowledge I have gained of field geology—the importance of fossils regarding our researches in stratigraphy.

While wending our way home in the dusk of the evening, after a hard day's work, we came across a man diligently engaged picking a deep hole in the black shales. On enquiry as to what he expected to find, he informed us that he was prospecting for coal. Mr. Richardson told him he would never find coal in those rocks for they are thousands of feet below the coal formation. But the next day when we passed the same place the man was still working away and had employed another man to assist him. A right knowledge, as I afterwards learned, of the various formations which constitute the "crust" of our earth and its fossil remains, would have told this man that coal could never be found in those rocks.

Coal occurs in many formations. It has been mined for over a hundred years in Beven, Sutherlandshire, in rocks of the Oolitic epoch, in other countries in the Miocene-Tertiary beds, and other rocks belonging to the upper crust of the earth, but the Carboniferous deposits are the coal bearing beds of Great Britain, the United States, Nova Scotia, Australia, and other countries.

The miner who has studied his geological text-books knows full well that it is utterly useless to look for coal in rocks containing Graptolites, trilobites and other organic remains which existed untold ages before the coal deposits were formed. I shall speak of other useless attempts to find coal in Lower Silurian and Cambrian rocks later on.

REMINISCENCES

During the same year (1863) I spent several weeks in the State of Vermont examining and collecting from the limestones along the shore of Lake Champlain, and along the road from Whitehall to Burlington, the whole of which distance I walked, staying over night and sometimes longer at the various villages or towns on the way.

Before starting on this journey several days were spent at Whitehall, and here Sir William Logan joined me and together we examined the various rock formations of that vicinity and collected fossils and interesting facts relating to or connected with Canadian geology.

There was a good hotel at Whitehall, and here Sir William and I occupied a double-bedded room. On the 18th of October, Sir William informed me that he would leave the following morning by the early train for Albany, where he was going to see Prof. James Hall. At five next morning I woke to find Sir William scraping the mud off his boots and doing his best not to wake me, but he did, and I asked him if I could assist him. "No," he said, but soon I was roused by his shaking my arm and saying "Wake up, Weston, and lend me some money; I haven't enough to pay for my railway ticket." I gave him what he needed, but told him I had a hundred miles to walk, and when I reached Burlington I should want money. He promised to send a cheque to the care of the postmaster.

AMONG THE ROCKS.

After leaving Whitehall I spent a few pleasant days at a farm house at Comstock Landing. The geological structure of this vicinity is very interesting, but I am not sufficiently conversant with the formation here to say definitely what the horizon is—the fossils are obscure but look like Calciferous forms.

While here I received a letter from Dr. Godfrey, of Montreal, stating that my wife had presented me with a son, news which caused me to hasten on with my journey. The following morning (Sunday) I shouldered my fishing basket (which I used for carrying specimens) and started on a day's tramp. It was past six in the evening when I arrived tired and hungry at a small village named Bread Port. Finding no hotel, I enquired for a boarding house and was told there was none, but Mrs. Brown down the street sometimes took in travellers. I went to Mrs. Brown, but she sent me to another house. They too refused to take me in, and, after wandering all over the village to get a vehicle, in which I failed, I met a man who looked a likely person to assist me. I stopped him and said: "You call this place Bread Port, I believe, sir?" "Yes," he said, "this is Bread Port." "Well," I said, "I think it is badly named, for I can neither get bread to eat nor can I get lodgings for the night." Perhaps I looked like a tramp, but a few words of explanation and this good Samaritan took me to his house, and when his people returned from church I was well provided for, and we spent a pleasant evening singing and playing.

The following morning I continued my journey and in course of time arrived at my destination, Burlington, where the postmaster handed me the cheque I had asked Sir William to send, and also gave me Sir William's letter to him, of which the following is a copy. I insert it to show how careful he was in money matters:

"At Prosper Hall, corner Merton St. and
Delaware Turnpike,
ALBANY, 19 October, 1863.

To the Postmaster,
Burlington, Vermont.

DEAR SIR,—I send you herewith a money letter for Mr. T. C. Weston of Montreal, who will call at the post office for it. I send also a letter signed by Mr. Weston that you may see his signature. Will you be so good as to deliever both to him when he calls. Mr. Weston has been with me lately to Whitehall and to Comstock Landing collecting fossils. He belongs to the Geological Survey of Canada, of which I am the director. By this information you can frame questions by means of which to identify him.

I am, dear Sir,
Very truly yours,
W. E. LOGAN.

I returned to Montreal a few days later and spent the remainder of the time till my next excursion in my usual work.

AMONG THE ROCKS.

As I shall often refer to the various members of the Geological staff I shall briefly mention those who composed the staff when I commenced my duties in the Geological Survey.

Our distinguished chief, Sir William Logan, whose kind acts and fatherly counsel remained forever dear to the hearts of all associated with him, was a Canadian, born of Scotch parents, in Montreal in 1798, educated at the High School of Edinburgh. He distinguished himself as a geologist in the South Wales coal fields, 1838. In 1841, Sir William—then Mr. Logan, visited the coal fields of Pennsylvania and Nova Scotia and then commenced his studies of the rocks of Canada, and was shortly afterwards appointed by the Government to form a geological survey of Canada. During the Paris Exhibition in 1858, he received the gold medal of honor, and was created a Knight of the Legion of Honor. He was knighted by the Queen in 1856. Poor health and a desire to spend the remaining years of his working life on his beloved Eastern Township rocks caused him to resign his connection with the Geological Survey of Canada in 1869. He was succeeded by Mr. Alfred R. C. Selwyn, who was recommended to him by Sir Roderick Murchison. After a wonderfully active life, during which time, as Sir William often said, "I never had time to get married," he died at the ripe age of 77 years, at his sister's house in Wales,

REMINISCENCES

22nd June, 1875. "And now he sleeps in the quiet churchyard of Llechryd between his brother Hart and his great friend and brother-in-law, Mr. A. L. Gower. Peace to his memory. Honour to his name."

Those interested in Logan's history and geological work I must refer to the Geological Survey and to the "Life of Sir William Logan, Kt., LL.D., F.R.S., F.G.S., etc., by Bernard J. Harrington, for a number of years chemist to the Survey, and now of McGill University, Montreal. I cannot close this page without recording a regret that in his memoirs of Logan's life Dr. Harrington omitted to even mention the name of the writer of these notes—who travelled thousands of miles with him, shared some of his hard field work, and was his almost constant companion in his last investigations of the Eastern Townships geology.

MURRAY.—Shortly after Logan's appointment, by the Provincial Government of Canada in 1842, to make a Geological Survey of Canada—as far as his means would permit—he secured, through the recommendation of Sir H. T. De la Beche, the services of Mr. Alexander Murray, a young man educated in the Royal Naval College of Portsmouth, and who served in the navy. Mr. Murray became a staunch friend of Sir William who trusted him with many important surveys. After many years faithful service—through the

Yours truly
W. E. Logan

recommendation of Sir William, Mr. Murray became director of the Geological Survey of Newfoundland, a position he held to within a short time of his death, in 1884 in his 75th year.

T. STERRY HUNT.—T. Sterry Hunt, afterwards Dr. T. Sterry Hunt, LL.D., Officer of the French Legion of Honor, etc., etc., succeeded Count De Rottermund—for a short time chemist to the Canadian Survey—in 1847. Dr. Hunt was one of the most able chemists and mineralogists of his day. His writings are known all over the scientific world. Dr. Hunt held his position as chemist and mineralogist to the Geological Survey of Canada for 25 years, and resigned this position in 1872, three years before Sir William's death.

BILLINGS.—Mr. E. Billings a Canadian lawyer was appointed Palæontologist to the Survey in 1856. Long before this Mr. Billings had been a zealous worker in palæontology, and had published many contributions to that science. These papers attracted Sir William's notice, hence his appointment. Mr. Billings remained a faithful worker in the Survey for 20 years, died in 1876, and was succeeded by Mr. J. F. Whiteaves.

RICHARDSON.—While on one of his geological surveys in 1846 Sir William met with Mr. James Richardson—a Scotchman who had seen much hard pioneer farm work in Canada, and had for a time taught in a country school in the Eastern

REMINISCENCES

Townships. Logan wanted a man to do camp work and engaged Richardson. After a short time, however, Mr. Richardson became so fascinated with geology that he induced Sir William to allow him to attempt a little geological work, so he was sent off to collect specimens and make notes of what rocks he saw. On his return to camp Sir William was so pleased with this Mr. Richardson's first geological work that he was again sent off to make measurements by pacing, which work proved so satisfactory that he was in a short time appointed explorer, etc., to the Survey. He became one of Sir William's most trusted field men, did a great amount of good geological work, as the reports of the Survey show, and was trusted with the arrangement of the geological specimens in the Paris, London, and other exhibitions. He remained a faithful worker in the Survey till after Sir William's death, was superannuated in 1879, and died three years later.

BARLOW.—Mr. Robert Barlow, formerly in charge of a corps of the Royal Engineers in a topographical survey in Great Britain—joined the Geological Survey in 1857 as chief draughtsman, which position he held until a short time before the Survey was removed to Ottawa. Mr. Barlow only lived two or three years after his retirement from the Survey.

SCOTT BARLOW.—Mr. Robert Barlow was ably assisted by his son Mr. Scott Barlow, whose chief work was to plot surveys from field note-books. In later years he did some excellent field geology in Nova Scotia, among the Carboniferous rocks. He was appointed chief draughtsman to the Survey department a short time before his death, which occurred in 1894 through blood poisoning.

SMITH.—Mr. Horace S. Smith, artist to the survey, was engaged in England, and came to Canada a year or two before my appointment in 1859. His duty was to draw fossils to illustrate the reports of the Survey. His drawings illustrate the Geology of Canada for 1863, and other Survey publications. Mr. Smith died a few years before the Survey was removed to Ottawa.

BELL.—Mr. Robert Bell—then in 1859, a young Civil Engineer—although engaged at various times in geological work, was not appointed to the Survey till some years later, since which time he has made many important surveys in various parts of Canada. At this date, 1897, Dr. Bell is actively engaged in geological work ; is assistant director, is M.D., C.M., LL.D., F.R.S., etc. Dr. Bell, with the exception of myself, is the only surviving member of the old regime of 1859.

I must not forget to mention here another permanent member of the staff—Michael O'Farrell, office keeper, whose connection with the Survey dated from its commencement to 1889, when he was superannuated. Michael was a thorough

Irishman, always ready to assist a friend in want, but a bitter foe to those who slighted or opposed him. His commanding appearance gained for him from some of the members of the staff the sobriquet—" Figure-head of the geological ship." Michael was a keen horseman, and had in his younger days ridden behind the hounds with some of the most popular noblemen of "dear old Ireland." Michael was very much attached to, and a constant attendant on his chief, Sir William Logan, whose disregard for personal adornment frequently raised Michael's ire. "Sir," said Michael to Sir William one day: "The tailor says these pants are not worth repairing." "Well," said Sir William, "confound him, tell him to make me another pair. "And your boots too, Sir, are in a very bad state." "Well, but," Sir William said, "can't they be patched? you know I can't walk in new boots."

Although Sir William did not forget Michael in his will, he never seemed the same man after his chief's death, and soon after the removal of the Survey from Montreal to Ottawa failed in health so much that it was necessary to appoint another janitor to look after the offices and cleaning of the museum; so, as I said before, Michael was superannuated, and was taken care of by his two daughters. Michael only survived about two years—after his retirement in 1889. He was succeeded by Mr. Thomas Burke, who at this

Very sincerely yours
T. Sterry Hunt

AMONG THE ROCKS.

date, 1897, is fulfilling his position to the satisfaction of all members of the staff.

Besides the small band of permanent workers spoken of, there were several other "extra hands," among whom was my old friend Mr. James Low, of Grenville, P. Q., who, though only a rough farmer, through accompanying Sir William on his explorations, became an expert in tracing certain bands of Laurentian rocks. His surveys were plotted by Mr Scott Barlow.

Another was Mr. Richard Oatey, a Cornish miner, who did much good work in the examination of mineral deposits. He was associated with the gold mining at the Chaudiere River, P. Q., at the time when the seven pound weight of gold was taken out of the beds of the rivers, and fields of that district.

Many other men were occasionally employed and rendered valuable service, but I can only mention here one other: Principal Dawson, of McGill University, Montreal, who, notwithstanding his arduous duties at the college, found time to assist Sir William in determining fossils, &c., &c. Sir J. W. Dawson, C.M.G., LL.D., F.R.S., F.G.S., &c., &c., retired from McGill University a short time ago and is now (1897) actively engaged with his books and fossils. His writings are known throughout the scientific world.

CHAPTER II.

EOZOON—"THE DAWN OF LIFE"—A REMARKABLE GEOLOGICAL CONTROVERSY—DISCOVERY OF TUBULI STRUCTURE—UNLIKELY THAT THE EXTENSIVE LAURENTIAN SYSTEM WOULD ONLY CONTAIN ONE KIND OF ORGANIC REMAINS—THAT THE HURONIAN SYSTEM HAS NOT YIELDED ANY FOSSILS.

THE great Eozoon controversy commenced soon after my appointment with the Survey (1859), has continued to the present day, and is likely to continue, with a gradual falling off of those who in the early days of its discovery favored the supposition that it was a fossil.

It would take many pages to record even a slight summary of the literature published for and against this remarkable organic looking structure. Here I can only give a slight outline of its discovery and a few remarks on the scientific interest and work done to prove this concretionary structure to be organic.

The first specimen of Eozoon known was found at Burgess, Ont., by Dr. Wilson, of Perth, Ont., and was sent to the Geological Survey as a mineral. Years after, it became associated with the supposed Laurentian fossil and was analyzed and examined microscopically, but nothing was found

AMONG THE ROCKS.

resembling organic matter. Later on, in 1860, other coralline-looking specimens were found by Mr. J. McMullen at the Grand Calumet, Ont. These specimens were sliced and examined under the microscope, but no fossiliferous structure found—a description of these can be found in the Geological Survey Report for 1863. They were given to Mr. Billings to study, and he finally determined not to risk his palæontological reputation in describing such specimens to be even probably of organic origin. These specimens were exhibited in London in 1862 by Logan, but he appears to have found few believers in these supposed Laurentian fossils.

About this time Mr. Low collected some fine blocks of serpentine intermixed with crystalline limestone (calcite), from Côte St. Pierre, P.Q. In these specimens Sir William noticed peculiar concretionary matter resembling the Black River fossil *Stromatopora rugosa* (Hall). The writer was at once requested to make microscopic sections of this rock. Several slides were made and the first looked at under the microscope revealed tubuli structure, or what is called the canal system, in the calcite part of the specimen, which is supposed to be the test or shell of the animal. Sir William examined this, the first discovery of what really resembled organic matter, with great interest, and, in his usual prompt way of acting, had Mr. Smith, the artist of the Survey, make drawings. These, with the micro-sections, Sir William took to Principal Dawson, who was delighted to find the tubuli structure in the calcite as he supposed it would be, and not in the serpentine, as Dr. Hunt supposed it would, if found, occur.

I have been particular in recording this the first specimen which appeared to have definite organic structure, because in his "Dawn of Life" Dr. Dawson omitted to give me credit for being the first to notice this structure. The illustration opposite p. 24 is from a steel engraving made from Smith's drawing, magnified about twenty diameters.

After this discovery a large portion of my time was taken up in microscopical work, examining and making thin slices of any Laurentian rock we thought might possibly contain Eozoon, and assisting Mr. Billings in palæontological work. In the summer of 1873 I was sent to inspect and collect specimens from our now celebrated Eozoon locality at Côte St. Pierre. Many beautiful specimens were found, some of which are now in the cases of the Geological Survey Museum; some are beautifully weathered, causing them to resemble the Black River fossil *Stromatopora* already spoken of. A short time after this, 1874, while collecting fossils in the rocks of the Guelph formation at Hespeler, Ont., I was telegraphed for to return to Montreal and accompany Principal

AMONG THE ROCKS,

Dawson to Côte St. Pierre. We arrived at Papineauville in the afternoon, took the stage to St. André d' Avelin (about twelve miles) and put up for the night at a small French boarding house. Here we could only get one bed, which Dr. Dawson took, while I lay on a rough couch, from which I was soon driven—well, not by mosquitoes. The following morning we continued our journey and arrived at Côte St. Pierre about noon. We found lodging in a small French-Canadian farm house within a few paces of the original spot where Mr. Low collected the fine blocks of serpentine in which the first specimen of tubuli structure was found.

During our journey from Papineauville we passed over great ridges of gneissose rock, serpentine, crystalline limestones, diorite, and other typical exposures of the great Laurentian system —over hills and through valleys with patches of cultivated land from which wheat and other crops had just been taken. Much of the land is covered with a second growth of young maples and other trees. A short distance from our Eozoon locality we pass over the beautiful Nation River, which winds through the hills and valleys of this picturesque country.

After a good meal of pork and beans, and a walk of a few paces, we stood on the slope of the hill in Low's excavation made in taking out the blocks of serpentine, in which, as I said, the first Eozoon tubuli structure was found. But besides this excavation we discovered that much work had been done by an American who was in search of "Cotton rock," a fibrous serpentine, or asbestos, now largely used in packing safes, etc. Here there is a thickness of about a hundred feet of the Eozoon-yielding rock exposed. It is associated with coarse diorite, gneiss, pyroxene, dolomite, tremolite, and other characteristic Laurentian rocks and minerals, forming high ridges and a very undulating country. The concretionary looking (Eozoon) portion of the serpentine rock, i.e., laminated serpentine and calcite, has been beautifully weathered. The soft calcite has been worn away and the narrower bands of serpentine remain standing up above the calcite, giving, as I have before remarked, a close resemblance to the Protozoon fossil *Stromatopora*.

In a short time we made a fine collection, some specimens of which are now historical, having been figured to illustrate the various writings on *Eozoon* and afterwards placed in the Geological Survey, and Redpath Museums. One of these specimens is reproduced in Dawson's "Dawn of Life," Plate III, and it was from this collection that the sections were made from which my micro-photographs were taken, some of which also illustrate the "Dawn of Life."

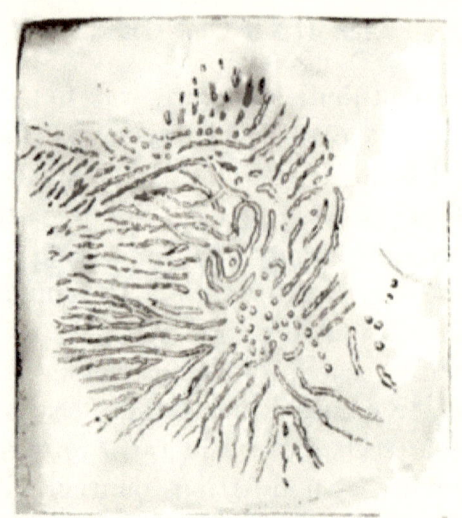

Eozoon Canadense — Canal System.

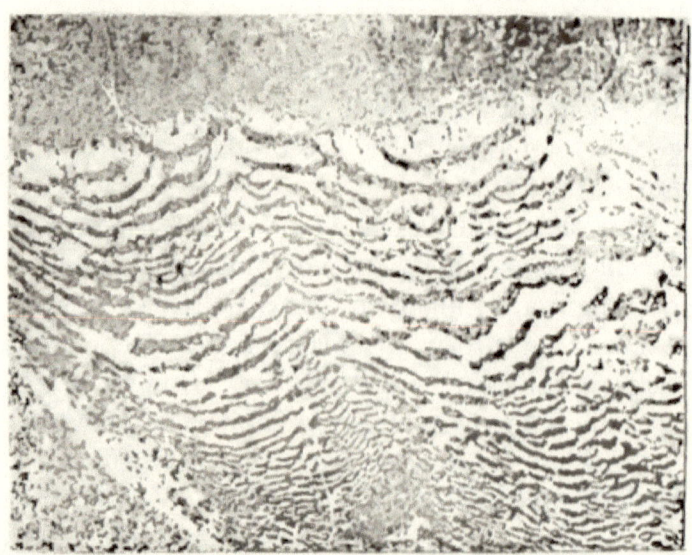

Eozoon Canadense: Laminated Structure.

AMONG THE ROCKS.

I must, at present, close this brief history of Eozoon—the gigantic Protozoa of the Laurentian seas, according to Sir J. W. Dawson; who after all these years and amid all the doubt of able palæontologists and naturalists, is to-day the firm believer in Eozoon he has ever been. For my part,—and I have made and examined many hundreds of specimens, both as thin sections for the microscope and decalcified specimens showing the tubuli structure,—I cannot say I am a believer in Eozoon being organic. No one, however, would be more pleased than I to see its organic nature established. This I fear will never be done unless something more definite than this supposed *Stromatopora*, *Stromatocerium*, as it is now called, is found.

The reproduction opposite p. 24 is from one of the etched specimens, and with the description, copied from the label on the large block of Eozoon serpentine in the Geological Museum, Ottawa, will give those who have not access to books and specimens a general idea of this supposed fossil.

"A gigantic Protozoon of the order *Foraminifera* from the Laurentian system.

This specimen is part of an unusually large mass of the fossil, contorted by the disturbance of the containing beds.

The laminæ of white calcite are the walls of the test of the animal, and the laminæ of green serpentine fill the spaces originally occupied by its gelatinous animal matter. Collected by Mr. James Low at Petite Nation, on the Ottawa."

When one considers that in Canada and British America the Laurentian system covers from 2,000 to 3,000 square miles and has in Canada a known thickness of 40,000 feet of strata, that part of this vast body of rock was originally sedimentary matter derived from older rocks of which we know nothing, and that other higher formations are almost wholly composed of the remains of animal matter, it seems strange that Eozoon should be the only thing that can in any way be attributed to organic structure. Whether Eozoon is a fossil or not, it will always remain an interesting subject of discussion for students in palæontology and mineralogy, and the specimens prepared by the writer for the London, Paris, Philadelphia, Chicago and other Exhibitions, together with the pile of literature on the subject remain forever as a proof of the deep interest taken in this supposed fossil Protozoa.

Like the Laurentian system, the Huronian formation, which consists of a thickness of about 20,000 feet of strata composed of quartzites, slates, limestone, jaspers, conglomerates and many other varieties of rocks, contains no definite fossils. Certain markings on slates and other rocks have been described and figured as the trails of mollusca, &c., but these markings are so indefinite that few palæontologists believe them to be of organic origin. In one of my papers on concretionary structure, published in the Transactions of the Nova Scotian Institute of Science, Series 2, Vol. I, I mention serious mistakes made in assigning to concretionary and other markings organic origin.

The hundreds of miles of Huronian rocks examined by the writer, both in Canada and Newfoundland, have yielded nothing which can definitely be regarded as organic. Billings' *Aspidella terranovica*, (Palæozoic fossils, Vol. II,) from the Huronian slates of Newfoundland certainly looks somewhat like a crushed and distorted chiton; but this also is one of those doubtful forms which will be looked at shyly by the palæontologist of the present day.

Thus in these two great systems—Huronian and Laurentian—no true fossils have been found. There are reasons why life should have existed in both the Laurentian and Huronian formations; but till definite fossils are brought to light, most palæontologists and geologists will, I think, consider these two great formations Archean or Azoic.

CHAPTER III.

A VISIT TO THE UNITED STATES—JOURNEY TO THE ISLAND OF ANTICOSTI—MY ASSISTANT PROVES FORGETFUL AND UNFORTUNATE—DESCRIPTION OF THE ISLAND—ITS GEOLOGICAL FORMATION—SOME QUEER INHABITANTS—I SIGHT A COUPLE OF BEARS—FURTHER MISFORTUNES OF JACKSON—DEPARTURE FOR QUEBEC.

MOST of the year 1864 was spent in assisting Mr. Billings in palæontological and Museum work. Most of the month of June was occupied at Troy, Bald Mountain and Glens Falls, N.Y., making collections of fossils from the Primordial and Trenton formations. These were needed for comparison with our "Quebec Group" fossils. The specimens obtained at Troy and Bald Mountain proved very interesting as they threw new light on the relation of certain rocks of N.Y. State to those of the "Quebec Group" of Canada.

At Troy I was entertained and accompanied in my researches by Mr. S. W. Ford, a young palæontologist of bright prospects.

At Bald Mountain, to be near my work, I sought and found board and lodging in a typical American farm house. The great kindness shown me in this hospitable home, the interesting nature of my work, together with the peaceful rest after the day's work will always cause this period to be remembered as one of the most pleasant of my early explorations.

On the 1st of June 1865, I received instructions from Sir William Logan to prepare for a journey to the island of Anticosti. I was to take a man who could act as cook and general assistant, the necessary camp equipment and provisions for two months. I was to examine and collect fossils from the coast rocks between English Head and Becscie River, on the south side of the island.

In a few days my arrangements were complete, and furnished with a Government official letter to the Captain of the S. S. Napoleon III, (which carried supplies to the Gulf shore lighthouses) I with my assistant started on what proved to be an exceedingly interesting trip, but which was frequently beset with laughable and often dangerous incidents, a few of which with a slight outline of the geological structure of the island I shall try and record.

I knew nothing of the man who was going to be my companion on a wild sea coast, excepting that my friend Mr. Root, who had recommended him to me, said he was a good cook and first-rate all round camp man. My first experience of any short-comings in Mr. Jackson (John Jackson was his name) occurred while waiting for the Montreal boat to start for Quebec. While standing near some of our baggage I saw a little woman

REMINISCENCES

bustling around looking anxiously for some one. Seeing me among a lot of boxes she came up and said :—

"Are you the man who is going to Anticosti?"

I replied that I was.

"Well sir," she said, "I hope you will take care of my husband, Mr. Jackson, for he can't take care of hisself, and please sir where is he now?"

"Oh, he has forgotten his overcoat," I replied, "and has run home for it."

"Ah, there it is," the woman said, "he never does remember anything."

Just at that moment Jackson came up and "All aboard" was called out. As we hustled the little woman off I gave her a few dollars on account of John's wages, and her last words to John were, "Don't forget the baby."

The delightful voyage down our noble St. Lawrence River, the courtesy of the officers on our good ship, together with the little trials and troubles of my assistant, will ever remain fresh in my memory. We reached English Head Lighthouse on the 18th of June, not, however, without several strong reminders of the "perils of the deep."

At the time of my visit to Anticosti in 1865, there were no inhabitants at the west end, excepting those connected with the lighthouse, but fishermen from other parts of the island frequently came on business to Mr. Malouin, the

AMONG THE ROCKS.

light-keeper. These men, like most fishermen, are a rough, hardy set, ever ready to assist one another in times of trouble, always ready to indulge in what they call little jokes, but, as will be seen by the following incident, very much opposed to having a joke played on them.

For several weeks Miss Malouin (daughter of the light-keeper) had been anxiously looking out for a priest, who at that time visited the island occasionally to administer religious functions, and as there was a probability of her confessor arriving that day she and I stood at the foot of the flag-staff speculating on his approach. In a little while we saw in the distance a boat containing several men, and with the Union Jack flying. In a moment all was ready to hoist the flag in honor of the priest's arrival. When, however, the boat came within a short distance of the shore, to the young lady's dismay she discovered that it only contained fishermen, who, knowing the priest was expected, thought they would play one of their little jokes by flying the flag, usually displayed in honor of an ecclesiastic. Seeing disappointment and anger depicted on the lady's face, I pointed to a fishing basket which lay near the flag staff; my meaning was instantly interpreted. "Good," said Miss Malouin; and in a moment up went the basket to the top of the pole. But, alas! my little joke was observed by the fishermen, and when they reached shore they made straight for me and were about to do me bodily harm, but the brave girl rushed in between us and cried, " Strike me if you want to beat any one!" on which the men slunk off, muttering threats of vengeance on my head.

At English Head, as along the greatest part of the south shore, a sloping reef extends seawards probably from half-a-mile to a mile. According to Bayfield, the outer edge of the reef forms a precipice from twenty to a hundred feet deep. Ships approach the edge of this shelving reef with great caution, sounding continually, and in case of fog standing off two or three miles from shore.

All lighthouse supplies are carried first in the ship's boats, then in a cart which is kept for that purpose; with these our supplies and camp outfit were landed. Letters of introduction to the light-keeper, Mr. Malouin, gained me a cordial welcome, and I was hospitably housed in the lighthouse till the busy time caused by the arrival of the Government supply boat was over, and I could get the cart to carry me to Gamache Bay, a distance, I think, of eight miles, where I decided to make my first camp.

During my short stay at English Head, the time was occupied collecting fossils and other natural history specimens. The reef, already referred to, which is dry at low tide, affords an excellent opportunity for collecting some of the characteristic fossils of the island, among which is the fossil to which Billings gave the generic name *Beatricea*, and classed it in the vegetable kingdom. Hyatt supposes it to belong to the Cephalopoda family, placing it with Miller in the animal kingdom. The latter supposes it to belong to the Bryozoa family, and assorts it with fossil sponges. After seeing many hundreds of these forms in *situ*, I quite agree with Billings in giving it a vegetable origin, and think it will ultimately be described as a fucoide. I have never seen a perfect specimen of *Beatricea*. The base or root of this plant-like form is frequently met with, but the upper end is always broken and does not show the termination, so at present we do not know what length this fossil attained. The longest section seen by me measured twenty feet and was about six inches in diameter, while fragments of others measured a foot in diameter They are cylindrical in form, slightly tapering towards the upper end; the exterior is rough, resembling the bark of some trees. A tube crossed transversely with highly concave septa runs the whole length, while between this central tube and the outside, concentric layers resembling those of an exogen tree and a coralline structure are seen. Billings describes these species from Anticosti as *Beatricea undulata* and *B. nodulosa* They occur in the Lower Silurian, and are most abundant in the Hudson River formation in the vicinity of the west-end lighthouse. This genus was found in the Hudson River rocks of Rabbit Island, L.S., by Dr. Bell, of the Canadian Survey, and by the writer, in 1884, at Stony Mountain, Manitoba.

My main object in visiting Anticosti was to collect fossils from the rocks between Macasty Bay, near the north-west end of the island, and Becscie River on the south side, a distance in a straight line of about thirty-seven miles, but very much longer taking in the numerous bends of the coast.

In these memoirs I can only speak in the briefest manner on the geology of that portion of the island visited; leaving those interested in the subject to peruse the able report of the late Mr. James Richardson, of the Canadian Geological Survey. The rocks of Anticosti belong to the lower and Middle Silurian. The various formations recognized by stratagraphy and fossils are: the Hudson River, Oneida, Medina, Clinton, and Niagara. A section of the Hudson River rocks at the west end of the island gives a thickness of 959 feet of strata, composed chiefly of argillaceous limestones, limestone conglomerates and other similar beds. These measures contain fossils characteristic of the Hudson River formation, a formation which occupies a great portion of the island, forming in places cliffs with an

*Yours Truly
James Richardson*

AMONG THE ROCKS.

elevation of from 100 to 400 feet, in places almost perpendicular.

Two days after our arrival at the west-end lighthouse we succeeded in getting the cart to carry us and part of our camp outfit to Gamache, or Ellis Bay, a distance of about eight miles. I sat with the driver, the light-keeper, while Jackson sat on our baggage. When within a mile or so of our destination, hearing Jackson whistling "Home, Sweet Home," I turned to ask him if he was thinking of "the baby," and glancing towards our baggage to my dismay I saw that several pieces of our camp equipment were missing. Fortunately, however, they belonged to Jackson. These things actually slipped from under him as we jolted along the rocky shore.

"Good gracious, Jackson," I exclaimed, "what's gone with your dunnage-bag, long boots and overcoat?"

"Good God!" Jackson said, jumped off the cart, and started back in search of his "duds."

Our camp was pitched, supper eaten and the shades of night were closing in when John returned with the lost baggage, which he had found just as the tide, which had risen, was about to cover them. John was in anything but an amiable humor, but a nip of rye from the medicine chest, and a good supper, soon restored peace in our camp, and before we retired for the night we bagged several large lobsters, one of which was the largest I have ever seen in Canada.

During our journey from the west-end we saw many seals, *Phoca Grœnlandica*, (Muller), which were very numerous at that time. They lay basking in the sun, but on our approach wriggled off into the sea.

Part of the coast between west-end lighthouse and Gamache Bay is exceedingly rough High escarpments of dark and light grey calcareous rocks—most of the beds of which are prolific in fossils, especially those of Junction Cliff, which, owing to the over-hanging beds, form good resting places for the fossils which had "weathered" out, and were gathered by handfuls.

Cape Henry on the west side of Gamache Bay has a vertical face of about 300 feet. Similar cliffs occupy a large portion of the coast, while intervening spaces are low land covered with a coarse grass. Many varieties of fruit-bearing trees and shrubs were seen between the west-end and Gamache Bay, among which were: the mountain-ash; gooseberry; currant; cranberry, &c. Pine, balsam, tamarack and other trees attain a good size, and much of this wood is fit for building purposes. During the ten days or so we camped in Gamache Bay we visited the only three people living in that vicinity: Luke, who kept the government store where a supply of provisions, &c., was kept, to be used only by wrecked sailors and others who had been cast ashore; an old man named Murray, whom we found in a log cabin at the head of the bay bedridden and in a terrible state of dirt and wretchedness, cursing heaven and earth, and all things on land and in the sea. So far as I could learn, this man was the only remaining member of a small band of wreckers, whose chief was the terrible pirate-king Gamache. The grave of this miscreant is situated at the end of the bay. If any of the descendants of Gamache's gang of pirates are living they could no doubt tell some fearful stories of wrecks caused by this execrable old wrecker. The only other occupant of the bay was a fisherman, who was away at the time, but his wife, who lived in a small log cabin, furnished us with lobsters, and other "fruits of the sea," and sheltered me for two or three nights when I was almost driven wild by mosquitoes and black flies. I was also indebted to this woman for the use of a boat to make a journey to River Becscie, a distance of about fifteen miles east of the bay. This journey was made to get a further supply of the beautiful bivalve fossil shell *Pentamerus Barrandi* (Billings), which was discovered by Mr. Richardson in 1856. Knowing this was to be a dangerous journey, after a good bit of coaxing and a promise to pay him $8.00 per day, I got Luke, who was well acquainted with the coast, to allow Jackson to mind his "wrecked sailor provisions store," and accompany me. We left the bay at 2 o'clock in the morning, when the tide was well up, but to avoid boulders and other obstructions we had to keep well out, perhaps a mile from shore. The moon was shining brightly, and as our frail boat glided through the lonely sea, I too, like Jackson, felt inclined to whistle " Home, Sweet Home."

At six o'clock we landed for breakfast and a nap, and while Luke prepared our meal I started out with my revolver to try and shoot a duck or two, which in that locality, were quite numerous; but before I had gone a hundred yards I found myself within fifty paces or so of two bears. They were separated from me by a small stream, and appeared to be feeding on dead fish. I gave a shrill whistle, a signal agreed on between Luke and myself, in case of danger. In a few moments he was at my side, and if I was surprised at seeing two bears so near me, I was more surprised to hear the blasphemous curses that Luke hurled at these denizens of the forest. Whether it was Luke's choice language that decided the bears to "get a move on" or not I don't know, but after surveying us for a moment they moved off towards the bush.

The next evening we reached Becscie River, where, after a long search, I found and made a good collection from the *Pentamerous* beds already spoken of. Owing to stormy weather we were detained here two days, and as the sea continued rough, and our small stock of provisions gave out, we decided to haul up our boat to a safe place and walk back. It was a tedious walk over those fifteen miles of rough shore, "A dreary beach with green waves tumbling free," but I was amply repaid by finding several new and important fossils.

During our absence, while keeping a sharp look-out on the Government store, Jackson had busied himself in administering to the wants of poor Murray, whose dirt and depravity had so shocked him that he was glad to be relieved from his task.

Jackson, who was an inveterate disciple of Izaak Walton, had, while we were away, discovered a fine trout stream, a short distance from the head of the bay. The fish were small, but very numerous, so I concluded to spend half a day fishing.

Early the next morning we set off for our trout stream, which I found as Jackson had represented, but here in the midst of our sport, misfortune again befell Jackson. While stepping on a boulder in the stream he fell flat in the water; his ducking did not, however, prevent him from continuing the sport. We reached camp that evening with sufficient fish to give all our friends in the bay a good feed.

The following day, according to previous arrangement, the horse and cart were brought to convey us again to the west-end. We remained there several days collecting fossils and other objects in natural history along the shore. Here, at the lighthouse work-shop poor Jackson's misfortunes followed him. We had obtained some leather, and John was busy mending a pair of boots, while I talked to the lighthouse keeper who was sitting on his work-bench. Suddenly the loud report of a gun immediately behind me; the shattering of a saw which hung on the wall near the light-keeper, and the exclamation "Mon Dieu" caused me to turn to where the report came from, and there I saw Jackson holding a large old-fashioned flint gun which had been tinkered into a cap-lock weapon. It appears this old gun had been loaded some time ago with shot and a ball to shoot a bear which had been seen prowling round the light buildings. However the bullet and some of the shot passed so close to Mr. Malouin that it grazed his coat before shattering the saw. Jackson had received a cut on his cheek by the nipple flying out, and his wrist and face were blackened with powder-marks which would remain as long as he lived. John was at once ordered to leave the buildings, and Malouin suggested that I put him in quarantine.

At last, after much trouble, I succeeded in getting a boat and men to take us to our last camping ground—Macasty Bay, a distance of

AMONG THE ROCKS.

about twelve miles, in a straight line from the west-end lighthouse, but as in our Beescie River trip we had to stand well out from the shore. During the journey fog set in, and that with a rolling sea, a small boat, four men, a large dog, which I had borrowed from the west-end lighthouse folks, eight boxes of fossils and provisions, rendered the journey a heavy and somewhat dangerous one, but at last we reached the bay and pitched our tent in a picturesque spot near a small stream of clear water and a considerable bush.

After a good supper of pork and pancakes, which latter Jackson was past master in making, the two men left us for the west-end. Jackson and I sat at our camp-fire till quite late and then turned into our blankets. How long we had been in the "arms of Murphy," as one of my men of former travels called it, I do not know, but I, always a light sleeper while at camp, was awakened by hearing some beast sniffing at our pork-box just outside the tent. I gave Tom, the dog, who was sleeping at my feet, a push, and hastily unhooked the tent door. Tom rushed out and gave chase to what I supposed to be a stray cow from the west-end lighthouse stables, but in the morning I was surprised to find the tracks of a large bear all round the tent. Tom slept on the outside of our tent for the rest of our camp there.

REMINISCENCES

We remained several days in Macasty Bay, during which time we made journeys in both directions along the cost.

Some of the cliffs in this locality are high and almost perpendicular, and a mountain rises to a height of about four hundred feet. This is an interesting spot for the palæontologist, for the rocks are very prolific in fossils, and besides the many new species of fossils obtained by us many more remain to be collected; indeed the field was so good for collecting that I had decided to remain a week or two longer and run the chance of another schooner other than the one which was to call for us. But on the following morning a schooner anchored opposite the bay, and the men seeing our camp came ashore. As I found they were bound for Gaspé, I decided, much to the delight of Jackson, who had an idea that at any rate his bones would be left on " the damned island," to leave. In a short time all our traps were on board and the boat was steered for Gaspé, but owing to head winds the journey lasted three times the duration it would have done with a fair wind. It was a constant tacking till we reached Gaspé Basin. The stench was so great in the cabin, and it looked such a delightful place for parasites, that I spent one whole night on deck. However, we reached Gaspé safely just as the steamer had left for Quebec. This obliged us to remain there a fortnight, which time was spent in exploring some of the interesting geological sections of that vicinity, one of which was at Douglastown River. Here considerable excitement prevailed for some time owing to the discovery of Petroleum springs. Oil is seen oozing from the sandstones which belong to the Lower Helderberg group (Upper Silurian). The few fossils found here were plants which have a Devonian aspect.

Another locality visited was Silver Brook, a tributary of the York River. Here petroleum is seen floating on the pools of water and oozing from the rocks.

At the time of my visit to this place (1865) boring for oil was in operation, but so far the wells had only yielded salt water and gas. Although this oil excitement is constantly cropping up, I am not aware, even at this date, more than thirty years after, that oil in paying quantities has been obtained. A few interesting fossils were obtained from the Silver Brook rocks, and samples of petroleum collected for the museum.

And now before closing this brief account of my Anticosti trip, and bidding good-bye to our friend Jackson, I must relate two other misfortunes which befell him during our stay at Gaspé.

Early one morning we provided ourselves with a lunch, and started for a long geological tramp.

REMINISCENCES

After a time we left the main road and took a bush path which followed the base of a range of high hills and cliffs. We had proceeded a mile or so when we came to a log cabin. Jackson, always of an enquiring mind, started towards the cabin, but when within a few feet of the door, a vicious bull-terrier rushed out and buried his teeth in John's leg. At that moment a woman ran out and with a stick beat the dog off. I demanded that the dog should be tied up, and when this was done examined the wounded leg. Fortunately the flesh was not torn. I bathed the limb with water from a stream close by and bound it with my handkerchief, and we started back. It was a painful walk for poor Jackson, and his leg was very much swollen when we reached the hotel, where a doctor attended to the leg. Jackson stoutly declined to have me send someone to shoot the dog, saying : " Perhaps it is the only protection that poor woman has in that lonely wood."

Like the proverbial cat with nine lives my servitor was soon able to be about again.

The steamer for Quebec was due in three days, so leaving John to look out for the arrival of our boat, and instructing him to have all our traps on the wharf in time, I drove off to inspect some supposed ore deposits and an oil spring. I was back half a day before the boat was due, but to my surprise she was in sight and making for the

AMONG THE ROCKS.

basin, but Jackson, where was he? I hurried to the hotel, paid my bill, and on enquiring for Jackson was told that he was, that morning, going down to the wharf with the two bottles of mineral water I had previously collected, and, taking a short cut down a steep embankment, had fallen, cutting himself severely with the broken bottles, and that he was then lying in the Harbormaster's office; there I found the poor fellow on a rough lounge, and covered with the Union Jack.

By this time the steamer was at the wharf. Going on board and explaining matters to the Captain, he kindly allowed his men to fetch our luggage from the store-room, and also to have Jackson carried to the ship. The pitiable sight my poor man presented will probably never be forgotten. We left Gaspé on the 27th of August, reached Quebec safely, and in a short time were on board our train bound for Montreal, which place we reached a week after leaving Gaspé. There I paid Jackson off and he at once started for his home, where I trust he found his wife and baby. I had long ago come to the conclusion that his wife was about right when she said: "He can't take care of hisself."

I never saw the poor fellow again, and think he died about a year after. So ends this brief account of our Anticosti trip. The remainder of the year 1865 and to July the following year, was occupied in labelling and arranging my Anticosti and other fossils and in my usual microscopical and museum work.

CHAPTER IV.

THE SOUTH PETITE NATION RIVER—THE TRENTON FORMATION—THE CALCIFEROUS ROCKS OF ONTARIO—WE LOSE OUR WAY IN THE BUSH—THE CHAZY FORMATION—A GUELPH FREETHINKER—AN ADVENTURE WITH A BULL—VISIT TO POINT LÉVIS—AN ODD MISTAKE AT OUR BOARDING-HOUSE—THE EOZOON CONTROVERSY.

ON the 27th of July 1866, I left for the South Petite Nation River and spent about ten days examining and collecting fossils from the Trenton formation. This series of rocks takes its name from Trenton, Oneida county, New York. Although its thickness in Canada does not exceed probably 1,000 feet, it is palæontologically one of the most interesting formations. It has been studied perhaps more than any other formation, and has yielded a vast number of species and genera; indeed the Trenton formation has been likened to a magnificent museum of well preserved shells. The extensive quarries of this limestone in Montreal, Ottawa and many other places have not only furnished stone for our principal buildings, but have supplied our museums with thousands of beautiful fossils.

In the distant days of the Geological Survey of Canada, of which I now write, the Government grant was small and one had to tramp many use-

AMONG THE ROCKS.

less (geologically) miles of country to save expense. It was on one of these journeys in the South Petite Nation district that I took what I supposed to be a short cut through the woods, but long before reaching the main road leading to the log cabin where I was quartered, night came on, and I had frequently to look at my compass by the light of a match. It was while groping for my road in this way that I stumbled over what I supposed to be a bear, but which I found by striking another light to be a cow, and to my delight I stepped into a path which led to the road. In a short time I was safely housed, and if it was a one-roomed log cabin, and if I did have to sleep in the one bed while Mike and his wife camped on the floor in close proximity, and the youngsters slept in another corner of the room, I was thankful for even that shelter and also for the "God bless you" I received, when after a hearty breakfast of milk and potatoes I shouldered my large fishing basket and started for fresh fields.

A short time after returning from the Petite Nation River I left for Chambly (a short distance from Montreal) and spent a few days on the Hudson River rocks. I then visited Kempville, Oxford and other localities in Ontario, and worked among the Calciferous rocks, collecting many interesting fossils among which were several new species.

REMINISCENCES

The Calciferous sandrock received its name from the New York geologists. It is developed in many parts of the United States, where, as in Canada it succeeds the Potsdam formation and is chiefly a magnesian limestone. Its thickness in Canada is supposed to be about 300 feet. It is not very prolific in fossils, nor are the species numerous as in the succeeding formations. It is supposed the Lamellibranchiata—that numerous class of bivalves which are not equilateral, or divided equally on either side of the apex—commenced their life in this part of the Lower Silurian.

Before the season ended, I accompanied Mr. Billings to Isle Bizard, where we examined the Calciferous and Chazy rocks. Our chief object was to collect specimens of that characteristic fossil of the Chazy formation, *Bolboporites*. The weather was lovely, and Mr. Billings, who seldom left the office for field work, was like a schoolboy let loose for a holiday.

Here a little incident caused us a good laugh.

We had finished our work and started on our return journey. Opposite the house where we had boarded was a considerable bush, and to save a long detour we determined to cross it. Blocks of Chazy and Calciferous limestone were strewn over the ground, and we kept going from one to another, picking up a fossil here and there, and working, as we supposed, towards our destination, but to our surprise, instead of being a mile or so away, we came plump up in front of the house we had left. I laughed heartily in which Mr. Billings joined at the time saying. "Take a bearing with the compass, and you lead the road." This time we got through the bush all right and were soon back again in Montreal.

The Chazy formation takes its name from Chazy, Clinton county, on the west side of Lake Champlain. It has fine beds of workable limestone, both for building and for ornamental work. Some of the beds at Montreal, Caughnawaga, and other localities furnish a good marble. It contains many species and genera of fossils, and is characterized by the small Brachiopod *Rhynchonella*, of which some of the beds are almost wholly composed.

Phosphatic nodules, which are supposed to be coprolites, occur in abundance in some of the beds; they are composed chiefly of the remains of the bivalve *Lingula*, which is supposed to have formed the food of the animals from which the coprolites were derived. It is in the Chazy formation that we find many interesting Crustacea (Trilobites), some of which are rolled up as if to protect themselves from danger, as some of the lower animals of the present day do.

My field work during the summer of 1867 was confined to collecting fossils from the Guelph formation at Guelph, Galt, Elora, Hespeler, and other localities where the Guelph rocks are exposed. This formation takes its name from the town of Guelph, on the river Speed, Ontario. It occupies a position next above the Niagara formation, and forms the top of the Middle Silurian: is about 160 feet thick and is composed chiefly of yellowish white dolomite, or magnesian limestone. Some of the beds are hard, and make a good building stone. The beds at Guelph and Galt are characterized by casts of the large bivalve, *Megalomus Canadensis* (Hall).

The Guelph formation has yielded a large fauna of fossils, many of which were described by the late Mr. Billings. Other new species are now being worked out by his successor, Mr. Whiteaves, the present palæontologist to the Geological Survey of Canada. The numerous fossils—many of which were new to science—now in the cases of the Dominion Geological Museum, show some of the result of our labor among these interesting rocks.

I close these brief remarks on the Guelph rocks by recording the name of my old friend Mr. James Harris of Hespeler at whose house I boarded. Both Mr. Harris and his two daughters worked in the woollen factory of Hespeler. They were humble people, and while with them I did not "live on the fat of the land," but they gave of the best they had. This was the first time I had lived with a confessed infidel—

for such was old man Harris—and what with my fossils and the learned talk of my host I found plenty of food, if not for the body, for the mind. There was one thing remarkable about the old man's belief, and that was, when Sunday came he always insisted on his children attending Sunday school, and learning their catechism.

Old man Harris is always associated in my mind with the Guelph formation. The old man must be dead long ago. I wonder whether he still holds the same views of religion. As I visited Hespeler and my old friend later on I shall perhaps speak of him again.

The following month and until June 17th, 1868, I was fully occupied with museum, microscopical and other work relating to the palæontological branch of the Survey. Then on the last named date commenced a series of geological excursions which proved highly interesting and furnished much new palæontological information regarding the great range of rocks on the south side of the St. Lawrence between Point Lévis and Gaspé, and now known as a portion of the "Quebec group," but which, owing to the lack of fossil evidence, were in the early days of the Survey assigned to various horizons.

My examination of these rocks commenced at Rivière Ouelle, 80 miles below Quebec, to which place I was sent by Sir William Logan to examine in *situ*, a series of black phosphatic no-

REMINISCENCES

dules, which were then, as now, supposed to be coprolites, and the singular cylindrical fossil-looking forms which occur with the nodules. These have been described in the Geology of Canada for 1863, and are spoken of by myself and illustrated by micro-drawings in the Geological Report for 1876-77, pp. 432-33. Part of the river shore is exceedingly rough and the beach is strewn, or was at that time, with the bones of the white whale, or Beluga, which were trapped here for their oil. Like the seal *Phoca Grœnlandica*, these small whales, or porpoise, as they are called by some, are of great antiquity, their skeletons having been found in the blue clay many feet below the surface, and far away from river or sea.

The cases of the Geological Museum, Ottawa, contain many fossil bones of the last two species, a great number of fossil shells and plants from the Post-Tertiary clays of Montreal, Ottawa and other localities. One case contains almost an entire skeleton of a harp seal—*Phoca Grœnlandica* (Müller). The label attached to this valuable specimen reads:

Skeleton of Harp Seal—

Phoca Grœnlandica (Müller),

Post-Tertiary clays, Montreal. Found at the depth of 30 feet below the surface in Peel and Conte's clay pit—1861. Weston.

AMONG THE ROCKS.

After collecting the above specimens I spent much time in freeing the bones from the clay in which they were imbedded, and then in articulating them, in which shape it remained till about 1887, when it was re-articulated and restored by Mr. Jules Baily of Montreal, thus, in my opinion, deteriorating the fossil aspect of the remains.

Besides the fossil bones and shells mentioned, these Saxicava clays (which are largely used for brick-making) contain, at Green's Creek, Gloucester, Ottawa, abundant nodules in which are found beautiful examples of *Mallotus villosus* (Cuvier); that delicate fish, the Capeling of the lower St. Lawrence, one of the few fishes which has weathered the storms and changes of centuries.

While pacing the distance along Rivière Ouelle and noting the various outcrops of rocks between the railway station and the coast, I had a laughable and what might have been a serious encounter with a bull. I was very intent on my task as this was my first attempt to make geological measurements with a view to plotting my work. All went well, till, seeing a peculiar clump of rocks in a field, I decided to make an offset from the river road. About half the distance was covered when to my surprise a wicked looking bull came tearing along from the other side of the field; for a moment my chance of escape seemed very slim indeed, but dropping my geological traps, I ran as one runs for his life, and succeeded in gaining the clump of rocks, up which I scrambled just as his bullship was about to hook me in the back. A shout brought a farmer from the next field, and after some little trouble I regained my instruments, but my line of measurement was never completed.

Soon after my return from Rivière Ouelle I was again summoned to Quebec to accompany and assist my old friend Mr. Richardson in his investigation of that portion of the "Quebec Group" occupying the coast between Rivière du Loup and Rimouski, a distance of about 60 miles.

During the autumn of 1867, and all the summer of 1868, Mr. Richardson was engaged in tracing out the structure of these Quebec rocks. Up to 1867, very little was known as to the palæontological history of the coast rocks between Rivière Ouelle, already referred to, and Rimouski. The discovery by Mr. Richardson this summer (1868), of one or more species of Trilobites was considered so important that he telegraphed the fact to Sir Wm. Logan and requested him to send me to assist in searching for other forms. I was at once despatched to Point Lévis, where I found Mr. Richardson waiting for me at the railway station. It was a dark night, and I was cold and hungry. In those days we had no luxurious parlor and dining cars, but in a few moments we were housed in the Victoria Hotel, that comfortable house which has since then sheltered Logan, Billings, Webster, Ells and other members of the geological staff, after days of hard tramps over the Lévis rocks.

A glance at the fossils Mr. Richardson had collected showed at once that they belonged to the Primordial formation, a fact which threw a new light on a large portion of the rocks of the St. Lawrence river.*

The following morning we bade good-bye to our genial hostess Mrs. Tofield,—who has long been numbered with the great majority, and whose comfortable resting place for the traveller has long been closed owing to the removal of the Grand Trunk, Quebec and Lévis depot to a more convenient site, in close proximity to the Quebec and Lévis ferry boats—and started for St. Roch, a small village, close to the shore of the St. Lawrence river, which place had yielded the precious fossils which were to aid so materially in working out the complicated structure of the Quebec rocks. Our first day's researches proved highly satisfactory, for we had bagged eight species of fossils most of which were new and all of a decidedly Primordial aspect. All these forms were obtained from the limestone pebbles of the conglomerates, a fact which left us to conclude that the matrix

---

*In these notes I speak of the various geological horizons as they were known then, and not as they are known at the present day, 1897.

REMINISCENCES

of the conglomerates and the other bedded rocks were newer and belonged to the Potsdam formation.

We continued our examination of these shore rocks at St. Anne, St. Denis, St. Paschal, Rivière du Loup, Trois Pistoles, St. Simon, Rimouski and other places. The result of our work was considered by Mr. Richardson very satisfactory.

Our mode of travel was chiefly by Grand Trunk Railway as far as Rivière du Loup; east of that to Rimouski we travelled either by buggy or by Her Majesty's mail (often being the whole night on the road) which sometimes consisted of a buck-board and one horse.

The construction of the Intercolonial Railway between Rivière du Loup and Halifax—a continuation of the Grand Trunk line—was then in progress, and afforded us both opportunities of travel and an insight into rocks hitherto concealed. During our journeys in this section of the country we saw much of the poor farming class of French Canadians. Few spoke even a little English, and as both Richardson's and my own French was very limited, we often made laughable mistakes; two examples I may mention here.

Being one day in a house where none of the inmates could speak English, I undertook while taking my soup to ask for a table napkin, but in some way mistook the French word "sheet" for that of "napkin." Judge of my surprise when the old lady brought me a large white sheet.

On another occasion our horse gave out and we were obliged to put up for the night at a log cabin where both food and accommodation were at a low ebb. My failure regarding the table napkin induced me to ask Richardson to try his hand at Canadian French, so when asked by the woman of the shanty " how many eggs we wanted for supper, and whether boiled or fried," which modes of cooking were conveyed to us by signs, Richardson said in French, as he thought, "four boiled eggs," but when we sat down to supper behold there were four dozen. We ate four and after assuring the poor woman that we would pay for the whole four dozen, we divided the remaining forty-four among the members of the family. They were still feasting on the remains of our supper when we entered the straw bed on the floor of the loft of the cabin. This trip finished my field work for 1868.

Between the last date to the following June I was fully occupied in the museum and in my workroom. The *Eozoon* controversy—of which I shall again speak later on—was still one of the interesting geological subjects of the day, and much of my time was spent preparing microscopic sections of Laurentian, Huronian and other rocks. Sir William Logan watched my operations with keen interest, and spent hours over my microscope examining the sections as they were prepared.

About this time Sir William appointed me Librarian to the Survey. I was to spend a short time every morning, before going to my more important duties, in recording the various publications purchased and presented, and attending to the distribution of Survey reports, etc. I retained the name of Librarian to the Survey for six years, when a Mr. White was appointed to relieve me from that work. The library at that time consisted principally of books belonging to Sir William, most of which were, after Sir William's death, purchased from the Logan estate. These books formed the nucleus of the library of the Geological Survey of Canada which to-day (1897) contains 12,000 volumes, and has for Librarian the learned and venerable Dr. J. Thorburn, who has two lady assistants. In the last Geological Report, 1894, it is recorded that during the year ending 31st December, there were distributed 5,666 copies of the Survey publication, This year (1897) 9,922 copies were distributed. This record of the doings of the library of the Geological Survey of to-day (1897) marks the great progress in this department since the time when I managed the affairs of the library in an hour or so each day.

CHAPTER V.

THE FIELD SEASON OF 1869—JOURNEY WITH MR. RICHARDSON—LOST IN THE BUSH—THE PRIMORDIAL ROCKS IN NEW BRUNSWICK—ARISAIG, NOVA SCOTIA, AND ITS COAST ROCKS—RETIREMENT OF SIR WILLIAM LOGAN—FIELD SEASON OF 1870—THE ROCKS OF QUEBEC—ANOTHER VISIT TO THE UNITED STATES—VISIT TO STANFOLD, QUE.—FIELD WORK OF 1871-2—THE COAST OF LABRADOR.

THE geological field season of 1869, was a busy and interesting one for me. Many hundreds of miles were covered and much information gained regarding the Primordial and other rocks then under discussion. The first journey made (June 12th) was to Belœil Mountain, P. Q., to examine a small outcrop of Devonian limestone. The distance between the small village of St. Hilaire and this limestone was measured by pacing and with a prismatic compass; the various varieties of diorite and other intrusive rocks which form the mountain were noted, and a small but instructive collection of fossils made from the limestones.

This mountain and exposure of limestone is always associated in my mind with my old friend and colleague Mr. James Richardson. It was the following summer I think when I again

REMINISCENCES

visited this locality with him to make further geological observations.

Arriving at the village already mentioned Mr. Richardson decided to cross the mountain through the bush while I took the cart-road to our limestone locality at which place we were to meet again. It was after one o'clock when I reached this spot. The day was hot and sultry and the mosquitoes thick and hungry. I was tired, and concluded not to wait for Richardson, but to take my lunch and sample the bottle of claret we had provided ourselves with from the village. After lunch and my after-dinner pipe I began to be anxious about my companion. But time wore on and the sun was near setting when I reluctantly decided he was lost—a strange conclusion to come to about a man who had spent a large part of his life in the bush. But coming to this conclusion I at once built a large smudge fire, and as the smoke ascended prayed that my dear old friend would see it. The shades of evening were just beginning to deepen when a rustling in the bush caused me to look up, and there was poor Richardson with clothes torn and bathed in perspiration. He sank down by the smudge fire thoroughly exhausted. I at once gave him a drink from our flask of claret, but for a time I feared he had a sunstroke, and was doubtful whether he could be got back to the village that night. After bathing him with water, however, which I carried in my hat from a spring some little distance off, and drying his clothes, he revived, and before dark we were on our way to the village. Mr. Richardson had followed a ridge of rocks for a long distance through the bush and, after several hours' tramp, found he had lost the bearing of the locality where he was to join me. Knowing I would wait for him at the appointed spot, he had struggled on, but at last saw the smoke of my smudge fire and was soon with me again. In after years when we sat by our camp fire he would often refer to this incident and say "I never was so exhausted in my life, and if I had not seen the smoke from your smudge, I should have lain down and perhaps never seen you again."

Part of June and till the latter part of July was spent in visiting various localities in New Brunswick. The Primordial rocks were occupying Mr. Billings' attention and it was thought advisable to have me examine and collect fossils from various localities where this formation was well developed.

Arriving at St. John I at once sought Mr. George Matthew of the Customs department, who, in conjunction with Prof. Baily, during their vacations, have done good geological work for the Survey. I wish to record here my thanks to Mr. Matthew for his kindness in furnishing me much information regarding the places I was

REMINISCENCES

about to visit, viz—Mouthpath, Ratcliff's Mill Stream, Clark's Point, Oak Bay and other localities.

The weather was beautiful, but occasionally foggy, as is frequently the case in the vicinity of the harbor of St. John. Prof. Baily joined me for a few hours, and together we examined and collected fossils from several of the exposures in and near the city. Quite a number of interesting forms were found, many of which are now in the cases of the Survey Museum. A journey was then made by buggy to Ratcliff's Mill Stream, a few miles from St. John, where the slates are prolific in Primordial fossils. It is the home of that pretty little characteristic Primordial Crustacean, *Microdiscus*—which I shall mention again while speaking of the Newfoundland rocks. Here I obtained lodgings with some of the members of the family after whom the stream was named. My bed was on the floor of the dining, sitting and general room. Outside a dense fog prevailed and the night was chilly, but in the house a wood fire blazed in a large open fireplace and I expected to sleep well after a hard day's work. Soon after lying down, however, it seemed that there were dozens of mice playing hide-and-seek, and when I did sleep I dreamed that these wretched little rodents were dancing a quadrille on the coverlet of my bed. In the morning, while I ate my porridge at one end of the table, a daughter of my host combed her raven tresses at the other end.

From St. John a journey was then made to Oak Bay, Clark's Point and several other places, chiefly in a stage coach much resembling the ancient English coaches which carried Her Majesty's mail and passengers before the days of railways.

Some of these excursions were exceedingly pleasant. From the top of one of these coaches one not only sees delightful verdant hills and valleys, lakes and streams, but occasionally gets a sniff of sea air, wafted across the hills and valleys from the Bay of Fundy.

My stay at Oak Bay was a pleasant one. Here, while collecting fossils, a man came up to me and asked many questions, among which were the same our Chinese friend of the yellow jacket, Li Hung Chang, has lately been asking so many of our country people, "Where do you come from?" "How old are you?" "What salary do you get?" In return for the information I gave him, he informed me that his name was Weston, and when I told him that was my name he was so tickled —as our American friends call it—that we became great chums, and it was through him I spent some most enjoyable evenings with the people of Oak Bay.

Having visited the localities mentioned and completed, as far as time would allow, the work allotted to me, I returned to St. John and was soon on my way back to Montreal.

On the 7th of August, a few days after my return from St. John, I was directed to proceed to Arisaig, Nova Scotia, and make a collection from the coast rocks of that place. About three weeks were spent at this work, and a large collection of fossils obtained, many of which were afterwards described by Mr. Billings. I shall refer to this locality again when speaking of my travels of 1873. The usual routine of museum, microscopical and other work, fully occupied my time till the following summer.

During this year (1869) two important events in the history of the Survey occurred—the retirement of our dear old chief, Sir William Logan, and the appointment of his successor, Mr. A. R. C. Selwyn. Logan had looked forward for years to the time when he would be able to lay aside the cares of the Survey and devote the remainder of his working days to his beloved Eastern Township rocks, but he continued to hold the reins of the Survey till this year, 1869, when he tendered his resignation to the Government, and during the month of December following Mr. A. R. C. Selwyn was appointed his successor. Sir William's health was then failing, but before he finally gave up field work, he spent many days working out certain facts relating to the geology of the Eastern Townships. It was my good fortune to be

AMONG THE ROCKS.

chosen to accompany and assist him in the last field work he did.

The following summer, 1870, was a busy one in field work. I had so far been very fortunate in fossil collecting; for this reason it was thought advisable to have further collections made and notes taken of any new cuttings or quarries in localities which had already been reported on. Receiving instructions from the director I started June 5th for Grimsby, Ont., where I was to make a collection of fossils and collect other geological information regarding the Niagara rocks. I was fortunate enough, at the pretty little village of Grimsby, to get board and lodging with Mr. Johnson Pettit, a veteran entomologist and collector of fossils, to whom the Geological Survey is indebted for many fine specimens of the latter.

The Niagara formation derives its name from Niagara, where, at the Falls, it is largely developed, consisting of shales and limestones. Many of the beds are very prolific in fossils, a great number of which have been figured and described by Hall, Billings, and other palæontologists. This formation covers a large area of Ontario (Upper Canada). After a short visit to Niagara Falls, Mr. Pettit and I set diligently to work. The large quarry, and constant wearing away of the softer beds of the cliffs afford a good opportunity to the collector of fossils. It was from the well-washed banks in the vicinity of the quarries of Grimsby that we obtained the fine specimens of *Caryocrinus*, and other forms now in the cases of the Dominion Museum.

Grimsby is a delightful spot, both for the palæontologist, artist, sportsman, and agriculturist, situated almost on the banks of Lake Erie, and within a few miles of Niagara—thunder of waters—Falls, which discharge 18,000,000 cubic feet of water per minute over an elevation of 150 feet of blueish-gray limestone. Hundreds of tons of grapes, apples, pears, peaches and other fruit are raised annually in this vicinity.

Johnson Pettit, of whom I have spoken, lived in the old Pettit homestead, a mile or so from the village of Grimsby. It was in this dear old farm house I spent many pleasant evenings and restful nights after the day's work was finished. The first night, however, was not a restful one. I had been allotted the spare bed-room which contained a large old-fashioned four-post bed. Immediately above the head of the bed were several pegs, and on one of these hung a large wide-brimmed beaver hat such as our Quaker great grandfathers might have worn. When I extinguished the light, the moon shone through the window giving the old hat and other garments an uncanny appearance. After a time I dozed off, and in a restless sleep dreamed that a skeleton had taken possession of the hat and,

AMONG THE ROCKS.

with it on his skull, was walking off. I jumped up and found the old hat had fallen from the peg. At breakfast I related the incident to the family and was informed that during the night there had been a shock of earthquake, which was no doubt the cause of this singular occurrence. The hat and other garments were removed from the room, and I slept peacefully the remaining nights of my visit.

A few days after my return from Grimsby I left for the island of Orleans, P.Q., to examine and collect fossils at St. Laurent village and other localities in that vicinity. It was at a place called St. Patrick's Hole, near the village of St. Laurent, that I was fortunate enough to find a good specimen of *Eophyton*. It was the first time this genus had been found in our Canadian rocks, and was considered important as it helped to determine the horizon of a portion of the rocks of the "Quebec group." There has been much discussion as to whether this fossil is a plant or the track of some animal. I quote the following remarks, made by Mr. Billings, from the Geological Survey Report for 1870-71 :

"On the south side of the island of Orleans, near the village of St. Laurent, *Eophyton* was found last spring by Mr. Weston. It is there associated with several species of fucoids identical with those that occur on Great Bell Island (Newfoundland). The rocks here referred to belong to the 'Quebec group,' but as they are considerably disturbed it is not impossible that they may be older and brought up by a fault."

REMINISCENCES

At any rate this genus was supposed to be typical of the Primordial zone, or not to occur higher than the Lower Potsdam formation.

Part of the month of July of this year (1870), was spent on the Chaudière and Famine Rivers, in the Province of Quebec. I have already spoken of the seven pound weight of gold taken from the Chaudière and its tributaries. Gold in the sands of the Chaudière has been known since 1835, from which time gold mining has been carried on more or less annually till the present time, (1897). In 1894 the gold deposits of the Province of Quebec yielded 1,622 ounces of the precious metal, valued at $9,196.

My work on these rocks was not connected with their gold deposits; but to try to find fossils, and obtain other information which would settle the disputed horizon of certain rocks of this vicinity. While at work here I had the pleasure of accompanying Mr. A. R. C. Selwyn, our new director, who was making his first geological explorations in Canada. Both Mr. Selwyn and I were, for a short time, the guests of Mr. W. P. Lockwood, superintendent of the Canadian and Northwest Land and Mining Company. To this gentleman Mr. Selwyn was indebted for much information about the auriferous deposits on the Chaudière and its tributaries; and to his charming wife, our hostess, we were both indebted for many kindnesses, one of which I must relate here.

AMONG THE ROCKS.

After taking breakfast one morning at Mr. Lockwood's hospitable house, I started out to examine the rocks of the Chaudière. When I returned in the evening I found my pocket-book containing fifty or sixty dollars was missing. Knowing that I must have lost it since leaving the house, I started to retrace my steps along the river. I had gone about a mile when I came to a small pool of water over which I had jumped in the morning, and there in this shallow puddle was my pocket-book saturated with water. I hastened back to the house, and was congratulated on the recovery of my money, and here the kindness of our hostess came in. She took the bank notes and other papers from the purse, spread them on plates and dried them in the kitchen oven.

Leaving the Chaudière, I next visited several limestone exposures on the Famine River, and obtained from known fossiliferous rocks many well defined fossils of a Devonian aspect. A list of some of the fossils of this locality is given in the Geology of Canada for 1863.

Most of the month of August was spent on the Trenton limestones of Winchester, Ont., and at the village of Industry, P.Q. A large collection of fossils was made and notes taken of any cuttings, etc.

During the month of September I again visited Whitehall, State of Vermont. The work done

REMINISCENCES

by Sir William Logan and myself in 1863 had interested Mr. Billings so much that he decided to visit the localities from which we had obtained fossils. We spent several days at this work and collected information which aided materially in connecting the geology of this part of the States with that of Canada.

We were working away at our fossil limestones, about a mile from the hotel, when Mr. Billings took a sudden desire to return to Montreal that day. Nothing I could say would induce him to stay longer. "All right," I said, "we will leave by the evening train, I have paid expenses so far, give me money for our return trip." "But," said Billings, "I haven't a dollar in my pocket." He was not well at the time, and this seemed to make him still more nervous. I at once wired the office to telegraph us funds to the Whitehall Canadian Express Company, but strange to say we received no reply. However I felt bound to get my superior officer home that night. "Give me your note-book," I said, and with that and my own I went to the express office, saw the manager, told him the fix we were in, and requested him to furnish us funds to return to Montreal, offering to leave our note-books as security till we refunded the money either to him or his company at Montreal. The old gentleman gave a hearty laugh and at once furnished the funds required, saying we could return it to

E. Billings April 1862.

his company at Montreal. I hastened to the hotel, paid our bill, handed Mr. Billings into a cab, and in a short time we were in our train bound for home. It is a rather singular coincidence that when I was in the same town with Sir William, in 1863, he too was without money and borrowed funds from me to pay his way to Albany as I have already related.

Before closing my remarks for 1870 I must record the death of my friend W. B. Hartley, a young engineer who joined the Survey three years, I think, before his death this year. Mr. Hartley was with Sir William Logan, engaged investigating the structure of the Pictou Coalfields, N.S. This work Hartley continued in 1870, but towards the end of the season overexertion and exposure in the mines brought on a sickness which soon terminated the life of this bright young man. Mr. Scott Barlow, of the Survey, attended him in his last sickness at the hotel in New Glasgow, N.S., and it was my painful duty to assist his father in arranging his papers and personal effects in his office at the Survey.

From the winter of 1870 to July 7th, 1871, I was occupied with the usual museum and other work of the Survey. On the latter date I went to Eaton, P.Q., to examine certain micaceous schists and other rocks in which no fossils had hitherto been found. Mr. Richardson, of the

REMINISCENCES

Survey, who accompanied me, left Eaton to attend to some private matters for Sir William Logan, but before going said, " You will never find fossils in these rocks." When he returned in the evening, however, I showed him several well defined corals, fossils of a Devonian aspect, imbedded in pieces of hydro-mica schist. For a moment Richardson seemed at a loss, but at last said : " Well, Weston, you ought to be knighted for this find." It may seem egotistical for me to mention this, but I do so to show the great importance Richardson, like Logan, attached to fossils.

On August 3rd, I was again in the United States, this time at Troy, N.Y., to which place I had been sent to ascertain whether the conglomerates of Troy belonged to the same horizon as the Bic rocks of the St. Lawrence.

I was assisted in this work by Mr. S. H. Ford of Troy (already mentioned), who was then studying the fossil fauna of that vicinity. We were very successful in obtaining fossils identical with those of Bic. Poor Ford was then just gaining fame as a palæontologist, but in a short time after I saw him last, his brain gave out, and his relations had to place him in an asylum.

On the 22nd of August, a few days after my return from Troy, N.Y., I left Montreal again for Bic ; this time to accompany Mr. Billings, who was then much interested in the fossil fauna of the Lower Potsdam formation. We visited several of the most interesting localities between Rivière du Loup and Bic. Mr. Billings was not well at the time, and as at Whitehall often became nervous and would not leave the house, preferring to remain indoors and study the fossils already collected. After a very interesting journey we returned to Quebec and spent several days in exploring the rocks of the Chaudière Falls. Billings was then studying the small bivalve fossil *Obolella*, and it was here we obtained much information relating to this genus.

I have mentioned before, that at this time we travelled between Riviere du Loup and Bic chiefly at night, with the mails which were generally conveyed from station to station by a buckboard and one horse.

Stopping in the night to wake up a sleepy post-office keeper, changing horses and waiting in the cold while the mail bags were gone through etc., rendered these journeys anything but pleasant. On one occasion a fat French woman got into our stage. It was midnight, and her head soon began to bob from one side to the other, and at last it came down with a whack on my shoulder. For some time that head rested there. But ah! those days have long passed, and one can now stretch in a luxurious sleeper, and arrive at his destination refreshed and ready for work.

REMINISCENCES

The next official journey was to Stanfold and other localities in the Province of Quebec. Sir William Logan still continued intent on his Eastern Township rocks, and it was at his suggestion that the Director of the Survey sent me to these localities. The winter of this year 1871, was a busy one for all hands in the Survey A perusal of the report of the Geological Survey for 1872-73 will show how varied my occupations were when I was not engaged in field work.

My field work, in 1872, commenced May 29th, when I again started out to continue geological investigations in the Eastern Townships. Hereford, Farnham and other localities were visited and additional information regarding the "Quebec Group" obtained.

I had been working at Farnham Centre two or three days, most of the time in one spot where I had found a very interesting set of fossils. During the day I had noticed people watching me, but keeping a safe distance away. When I returned to the boarding-house I heard the folks talking of a crazy man who had arrived there and was digging deep holes and breaking rocks in a field where he expected to find gold. I took no notice, but after supper was surprised to hear that a deputation consisting of the chief men of the village was waiting outside to interview me. I went out and found several men, whose questions led me to think they took me to be a "little off." But when I told them it was not gold I was after, but fossils, they seemed more convinced than ever that I was not "all there," and determined to visit me at my work the following day, but bright and early next morning I was away to fresh fields. Whether these men ever learned what a fossil is I do not know.

The fossil fauna of the "Bic Conglomerates" had proved so interesting that Mr. Billings requested the Director of the Survey to send me to the straits of Belle Isle where the Lower Potsdam Group is in an undisturbed condition. Sir William Logan who still took a keen interest in the working of the Survey, thought valuable information might be obtained there, relating to the "Quebec group." So on the 29th of May, I received instructions to prepare for a journey to the coast of Labrador. As assistant, I decided to take Mr. Willimott, a young Englishman who had recently been engaged to assist me in museum work. A few additions to my Anticosti camp equipment were made, and two months' provisions, packed, and we were ready to start.

We reached Quebec the following morning and boarded the lighthouse supply S.S. Napoleon III., whose captain had received government instructions to land me and my assistant at L'Anse au Loup, or some convenient point on the coast of Labrador.

During the landing of supplies I had an opportunity of again spending a short time on the island of Anticosti near the lighthouses, and also at two of the lighthouses at Newfoundland. While near the straits of Belle Isle we encountered severe storms, and were in close proximity to several icebergs ; fogs detained us for a day or more. However, we landed safely and pitched our tent at L'Anse au Loup, where we collected a large number of beautiful specimens of *Archeocyathus*, the first specimens of which were collected by Mr. J. Richardson before 1863. All Mr. Richardson's collection, except the one figured in the Geology of Canada for the latter date, were lost at sea together with the schooner on which they were shipped. The specimen referred to, he fortunately had in his pocket.

While the rocks of the island of Anticosti contain an almost endless variety of organic remains, the rocks of the coast of Labrador possess a very small fauna of fossils—records of past ages, when the scanty life of this planet consisted of a few simple seaweeds, a few sponges, a few different species of bivalves, and a few varieties of trilobites—animals belonging to the Crustacean family. No well defined plants existed, and no vertebrate animal had yet appeared on the face of the earth. Still the same sun which shines now shone then, and rain fell in those days as it does now. This is known by the mud cracks we find on some of the argilla-

AMONG THE ROCKS.

ceous, or clay rocks, as one sees at the present day on the clay of a dried-up mud-puddle, and as we often see on clay, pits made by rain drops, so we find them in the rocks we are speaking of.

While the island of Anticosti abounds in trees, some of a large growth, and abundant vegetation of endless variety—mile after mile of the Labrador coast is covered only by lichen and scrubby brush, the roots of which we used for our camp fire. Returning to our camp one day we were fortunate enough to find, a mile or so away, a lot of small pieces of sawn pine which may have drifted hundreds of miles. This we gathered up carefully, carried to camp, and used for kindling wood.

It would take pages to narrate all the incidents of this trip one would like to record; but I must bring this journey to a close. Towards the end of July a small steamer, "The Beaver," which was built in Scotland, and was sent out here for the fishery business, put into L'Anse au Loup for coal. I interviewed the captain and requested him to take us to Quebec, as I did not wish to run the risk of remaining on that dreary coast, probably till the fall. After much wrangling about the matter he consented to take us, providing I paid him £10, used our "own grub," and asked no questions. This I consented to do, and in a short time we were on our way up the St. Lawrence River. The captain seemed to

REMINISCENCES

have been drinking while on shore, and his first mate acted like a typical nigger driver. Such a journey as we had I never wish to experience again, but after a most dangerous passage we reached Quebec, and we left that ship as quickly as possible, and arrived the following day at the Survey with a fine, but hard-earned, collection of fossils.

During the month of August I spent ten days at Paquette's Rapids on the Ottawa River, making a collection of Black River fossils, and on the 1st of October I again went to Arisaig, N.S., and returned late in the same month with new information and another beautiful lot of specimens.

CHAPTER VI.

LORD DUFFERIN VISITS THE SURVEY—I ASSIST SIR WILLIAM LOGAN IN THE EASTERN TOWNSHIPS—ANOTHER VISIT TO ARISAIG—SOME OF THE PIONEER PHOTOGRAPHS OF THE SURVEY—A STORM OFF CAPE GASPÉ—A VISIT TO NEWFOUNDLAND—LAKE TEMISCOUATA— WORK AT HESPELER, ONT.—THE CONGLOMERATES OF BIC—DEATH OF SIR WILLIAM LOGAN—THE PHILADELPHIA EXHIBITION—THE EASTERN TOWNSHIPS AGAIN—DISTRIBUTION OF ROCKS AND MINERALS—ETHNOLOGICAL BRANCH OF THE SURVEY.

DURING the latter part of the year 1872, and the early part of 1873, several distinguished government officials visited the Geological Survey, among whom was Lord Dufferin, who had recently been appointed Governor-General of Canada. On his first visit to our museum Sir William Logan requested me to accompany His Excellency, himself and Principal Dawson, through the museum, that I might answer any question relating to my work. Sir William then occupied a small room on the palæontological flat as sitting, and sometimes sleeping room. The wall on one side of this room was covered wholly with large framed slabs of Potsdam Sandstone on which are well defined Crustacean (?) tracks *Climactichnites*, so named by Logan because they resemble a rope ladder. While Sir William was pointing out the beauty of this specimen, Dufferin said: 'That reminds me, when I was at the skating rink the other day I saw some peculiar tracks on the ice. I watched the skaters for some time to find out what made those tracks, and at last, would you believe it, I discovered they were tracks made by a lady's petticoats."

By June 12th, I was again in the field assisting Sir William Logan with his Eastern Township work.

Sir William's last days of geological exploration were fast approaching. Some of his stratigraphical work in the Eastern Townships had been disputed by Dr. Hunt, and feeling convinced that he was correct, Sir William labored strenuously to prove this. It was my privilege to be allowed to assist him. We labored together at Danville, P.Q. One day Sir William took me to some plumbaginous limestones and other rocks on the Nicolet River. "These," he said, "are the rocks Hunt wants to make Huronian. Now if we can only find fossils here, it will not only be a feather in your cap, but it will save my reputation as far as these rocks are concerned."

We had labored all day, going over the exposures, sometimes on our hands and knees closely scrutinizing them with our lenses hoping to find some weathered-out fossil. In the evening we were returning to our hotel without a fossil, when I saw a thin band of limestone, less altered than the rest, between the plumbaginous beds.

"Stop," I said, "let us try these rocks." I broke off a piece of the limestone, looked at it carefully with my lens, then looked up at Sir William and said, "I will bet $10 this limestone is full of fossils." Sir William took the piece of rock, looked at it with his lens, shook his head and said, "I don't see the ghost of a fossil." "Neither do I," said I, "but we will see what my microscope says about it." Sir William took the specimen, wrapped it in his handkerchief, and said: "I am going in to-morrow to Cartier's funeral, and will put this on your table in your work-room." I went back to the Survey some days later, found the specimen on my table, went to work at it immediately, and in half an hour or so ran down to Sir William and said: "I was right; that Nicolet limestone is made up of fragments of fossils, and they look so much like Trenton or Chazy forms, that I can't tell the difference." If ever I saw our dear old chief's face radiant with joy, it was then. This settled a question of grave importance to Sir William

Soon we were again on the Nicolet River. This time I took my camera and photo chemicals—there were no dry plates then—to take several views of the rocks. Two of them were reproduced for Dr. Ells' report—"Geology of Canada for 1886."

Sir William remained several days with me on the Nicolet River and took a deep interest in the work. Once when he was some fifty yards away from me, I shouted "another fossil, Sir William," To reach me quickly he waded through water which ran over his boot-tops. Soon after this Sir William made me a present of a microscope, whether in recognition of my work on the Nicolet or not I don't know. This microscope was purchased from Mr. J. F. Whiteaves, of the Natural History Museum, Montreal.

I have dwelt a little long on these incidents of the Nicolet River because I wish to record here the great importance Sir William attached to fossils, and because these were almost the last days of his field work.

On the 20th of July, I again left for Arisaig, Nova Scotia. A nervous feeling came over me when I started, which increased all the way, and when I arrived at Arisaig, after a thirty mile buggy drive from New Glasgow, I was in a high fever. My old friends, the McDonalds, came out to meet me with a welcome always extended to a visitor to that rough sea coast. I told them I was sick and would return when the horse was rested. The old lady hustled me off to bed in quick time, took my purse, and when the horse and man were rested and fed, paid and started them back. They gave me some of their simple medicines, for there was no doctor within twenty miles. I was delirious for a short time, but in three days was up and at work again. While I

AMONG THE ROCKS.

lay in bed Mrs. McDonald came into my room and smoked her black pipe, and her daughters would sit around in their bare feet, "keeping me company"—poor simple country girls. I often think of those days when these kind folks walked miles to get me better food than they could afford to eat. The old folks are still alive, but have long passed the allotted three score and ten.

I had taken my camera and chemicals with me, and my first work was to take a series of twenty-one photographs of the coast rocks. In this work I was assisted by young Archie McDonald. Some of these views have been reproduced to accompany Mr. Hugh Fletcher's report of the geology of this part of Nova Scotia (Geological Survey of Canada for 1882–1886.) A section of the rocks at Joseph McDonald's cove, by the writer accompanies the photo of these exposures. I feel proud of these photographs, as I may say they, and my Nicolet River views were almost the pioneer photographs of the Geological Survey of Canada. Years after when the dry plate process was instituted field photographs became so simple that few of the explorers of the Survey went out without a camera and plates. The collection of views taken in the field by the various members of the Survey numbers many hundreds and is of historical interest. Those of Dr. George Dawson, Bell, Tyrrell, McConnell, Dowling and myself give one an idea of the wonderful rocks and charming scenery of the great North-west Territories. Those of J. B. Tyrrell show us the immense herds of reindeer which roam over the plains of the great barren lands; those of Low the dreary lands of Labrador.

A great number of the fossils collected from the coast rocks of Arisaig have been described and figured by the late E. Billings, "Geology of Canada, Palæozic fossils 1874."

My last official trip of this year, 1873, was to Gaspé where I went to examine and collect from various fossiliferous beds, known as the Gaspé limestone, or Lower Helderberg group. The thickness of rocks exposed at Gaspé is said to be 9000 feet, the greatest part of which are limestones holding a large and varied fauna of fossils. Most of my stay here was spent a short distance from Cape Gaspé, where I was fortunate enough to get lodging in a fisherman's house close to the shore. Fish formed the principal food of these toilers of the sea. Fish for breakfast, fish for dinner, fish for supper ; but the invigorating sea breezes made me hungry enough to eat fried shark.

During my stay here a terrific storm swept twenty fishing boats from their moorings. It was a grand sight to see the mighty waves lashing the great cliffs. I had been invited to take a day off from work and accompany a picnic party to Gaspé Basin. Soon after we arrived at our camping grounds, the great storm came on. The only shelter we could get was in a log cabin which was soon filled with the women and girls of our party. We men occupied the barn and slept in the hay loft, where we spent a most unpleasant night. The following afternoon the sea had calmed down a little, and we started for Cape Gaspé, but soon five or six of the womenfolk were lying at the bottom of the boat. The sea was very rough but we had a good boat and men who had spent the greatest part of their lives boating. At length we reached the wharf, and found all the folk of that vicinity ready to render any assistance in landing us, which was no easy task, for as a large wave carried our boat alongside the wharf only one person could be hoisted ashore. I was grabbed by brawny hands and hoisted up with as little ceremony as though I were a coil of rope. At last we were all landed and many a fervent thanksgiving was uttered for our safe delivery from the perils of the deep. I returned to Montreal on the 13th of September, after an instructive and profitable journey.

The winter of 1873-74, was occupied as usual in museum and room work, preparing a number of microscopic sections of fossil woods to be reported on by Dr. Dawson, developing and preparing fossils for the museum and for study.

Sir William Logan was ever ready to assist his colleague of the pioneer days of the Canadian Survey, Mr. Alexander Murray, then director of

REMINISCENCES

the Geological Survey of Newfoundland, and when in the spring of 1874, this gentleman wrote to his old chief Logan, saying he was doubtful about the horizon of some of his rocks, and desired him, if possible, to send Weston to try and discover fossils in these rocks, which if obtained would settle the, to him, important question, I was requested by Sir William who had recently retired from the directorship of the Survey, but still continued to act as adviser, to ask for two months' leave of absence that I might proceed to Newfoundland. The director of the Canadian Survey, Mr. A. R. C. Selwyn, acting on Logan's advice readily granted the leave asked for, and I left by the next English mail steamer, which called at St. John's. Our ship arrived safely, Murray met me at the wharf, and I was soon housed in the best hotel in the city. Two days after, Murray came with a trap and an Indian named John, who was to act as guide and general assistant. We drove off through a country, which according to Murray, " was made of the chippings of the world," and at last arrived at Manuel's River, twelve miles from the city.

The following extracts from Transactions of the Nova Scotian Institution of Science—Vol. IX., Season 1895-96, will explain more fully the importance of fossils in geological investigations :—

AMONG THE ROCKS.

THE PRIMORDIAL SILURIAN OF NEWFOUNDLAND AND CANADA.

"In spite of the oft-repeated assertion of Professor Jukes and the late director of the Geological Survey of Canada,—'If the fossils don't agree with the stratigraphy; so much the worse for the fossils,' my long experience as a collector of fossils and close observer of the various geologic horizons leads me to think that if the stratigraphy does not agree with the fossils; so much the worse for the stratigraphy. To illustrate the faith the late director of the Newfoundland Survey had in palæontological evidence, I will relate one incident out of many similar ones known to the writer:—In the summer of 1874 Murray wrote to Sir W. E. Logan, then adviser to the Canadian Survey, saying: 'I have made my Manuel's River rocks Primordial; I am doubtful, however, whether my stratigraphy is correct; neither Howley nor I have been able to find the ghost of a fossil; could you arrange in any way to send Weston down for a few weeks.' The result was that I left by the next steamer which called at Newfoundland, and a few days after my arrival at St. John's was taken by Murray to Manuel's River where he got lodgings for myself and Indian guide. The following day I commenced my search for fossils, and in a short time was rewarded by finding, in the gray argilites, the well-known Crustacean. *Microdiscus Dawsoni*, (Hartt); which occurs in abundance in the Primordial slates of St. John, at Ratcliffe's Mill Stream, and other localities in New Brunswick. This Crustacean, *Microdiscus*, is a puny thing, not larger than the half of a small pea, but it told me a big tale about the geological horizon—told me that Murray's stratigraphy was correct, and that I stood on Primordial strata similar to those of St. John, New Brunswick."

I may mention here that the term Primordial, used by Barrande and the late palæontologist of the Canadian Survey, Mr. E. Billings, is seldom used now; St. John Group being thought a better name for that extensive group of rocks. This

REMINISCENCES

Cambrian division of the lower Silurian of Newfoundland, according to Murray, would, if found consecutive at any one locality, represent a thickness of 6,000 feet of black, grey and other colored argillites, micaceous calcareous slates and limestones, sandstones, conglomerates and other rocks, some of which are prolific in fossils, especially the iron-stained argillites of Manuel's River and other localities in Conception Bay. The fauna is similar to that of the Primordial of St. John, Ratcliff's Mill Stream and other localities in New Brunswick.

Mr. Murray procured lodging and board for us at a comfortable farm house a short distance from Manuel's River, a mile or two from Conception Bay. Before leaving us Mr. Murray said, "Don't let Joe get even a smell of fire-water," so when we were alone I said, "Joe, have you any fire-water?"

"No, master," he replied.

I decided before starting to work next day I would look round myself. Leaving Joe at the house, I remained away about three hours; when I returned I found Joe so drunk that he did not know who he was or where he came from. I searched his bag and found one full flask of whiskey and an empty bottle. These I took outside and smashed on the rocks. The old lady got poor Joe to bed, and the following morning my brave Indian was all right, and when I told

AMONG THE ROCKS.

him what I had done with his fire-water he said, "much better."

After breakfast, with some lunch in our pockets, we started off for the day. Murray and his assistant had searched the rocks of Manuel's River for fossils but failed to discover any. I was more fortunate, for the first day, not long after we started down the river, I found, in an exposure of clay-slate, one solitary specimen of that little fossil Crustacean, *Microdiscus*, already alluded to. Imagine my delight when I saw this wee fossil which, being typical of the Primordial fauna of St. John, New Brunswick, proved that Murray's stratigraphy was correct, and my journey to Newfoundland had been successful. But there was more luck in store. Some distance down the river I saw a small island with rocks in the middle of the stream. Joe found we could reach this by wading and ordered me to get on his back, which I did, and was soon standing on an exposure of iron-stained clay-slates crowded with fossils—*Paradoxides* and other typical Primordial forms. We collected specimens enough to fill several cases in Mr. Murray's museum. Two days after this was Sunday, and knowing how delighted Murray would be to hear of my discovery, I decided to start to St. John's after breakfast, but judge of my surprise when the folks of the house would not lend me a horse. "No," said the old lady, "Ye shall no brak the

Sabbath with one of my horses." But go to the city I would. Joe was quite willing to accompany with me, so, with some lunch in our pockets, we started off on our twelve miles' walk. The day was very hot, and when I arrived at Murray's house he was astonished to find we had walked in.

Murray was a kind-hearted man, ever ready to assist those in distress, but was noted for fits of temper, when he would use the strongest of "swear-words," and on this occasion he did use swear-words, and fairly boiled over with wrath against the old woman of our boarding house. But in a moment his anger was over, and, with a serious face, he looked up and said, "But what about the rocks?"

When I told him the Manuel's River rocks were crowded with Primordial fossils he was astonished and delighted, but wondered how he and that ———— Howley had missed them.

After returning from Manuel's River, Murray and I made several trips to various parts of Conception Bay, and obtained many interesting new fossils. I afterwards spent a short time on Bell Island, sleeping one night over the one living room of the house, where fowl, and a litter of pigs formed part of the family group.

Having completed the field work assigned me, I spent the remainder of the time till my steamer arrived in putting some of the cases in the small museum—which consisted of several rooms in

Yours very faithfully
Alex. Murray

AMONG THE ROCKS.

Mr. Murray's house—in order, and also in classifying the books in his library, during which work I found four Church of England prayer books. When I asked Murray how I would class them, he was very angry that any one should have "put their bibles in his library." He flung them across the room and said some of his choice swear-words. The next morning I went to the cathedral church, Mr. Murray was there deeply engaged in the devotions of the day.

Before leaving for home Mr. Murray, after paying all my expenses, insisted on presenting me with £10, and on my return to the office Sir William said "I have heard of your success, and wish you to accept $50.00 from me." This ended one of the most delightful trips I have had since my connection with the Survey.

A short time after returning from Newfoundland I was instructed to proceed to Lake Temiscouata and if possible obtain fossils from the conglomerates of that vicinity. Mr. A. H. Foord, who had lately been appointed artist to the Survey, accompanied me. It was his first geological excursion; the country was new to him and the scenery delighted his artistic eye. We got board and lodging at a French Canadian's house close to the lake, but poor Foord, who had been brought up in the most tender and delicate way by his English parents—suffered very much

REMINISCENCES

with bugs. Four times during the night he came to my room to report the doings of these pests. In the morning I found him asleep on the floor tightly coiled up in a sheet.

The rocks we wanted to see were on the other side of the lake, and the only boat I could get was a dug-out canoe, but I secured an experienced half-breed to take us over. The lake was a little rough, but we landed safely on the opposite shore. The wind however, continued to blow, and by the time the man was to return for us the water was too rough for his frail canoe. Darkness came on and relieved us from the ten thousand mosquitoes which my companion found even worse than the bugs, for this was his first experience of our Canadian mosquito. As bad luck would have it I had forgotten both pipe and matches—things Foord never carried—so we were unable to make a fire. I lay down under a tree and got a snooze, but Foord walked the beach all night, and at six o'clock the next morning, the water having calmed down, our man came for us. We were hungry, but none the worse for our night out on the lake shore. Foord refused to cross the water again so I left him to make sketches while I continued to explore the rocks of the lake. We returned home in about ten days, and after a short time I was again working at the Guelph formation at Hespeler, of which I have already spoken. Mr.

AMONG THE ROCKS.

Billings was then much interested in the fossil bivalve *Trimerella*, figures of which will be found in "Palæozoic Fossil Vol. 1. 1861-65." It was for the purpose of getting further information of this genus, together with other new fossils, I made these journeys to Hespeler, Galt, Guelph, Elora and other places.

While at Hespeler, on the 6th of September, I was called home to accompany Dr. Dawson, Principal of McGill University, to the now worldwide known *Eozoon* locality at Côte St. Pierre near the Petite Nation River, province of Quebec; the home of our supposed Laurentian fossil.

In the early pages of these memoirs I have spoken of this supposed organism of which page after page might be written; but I have neither space nor inclination to continue this subject, and shall only add here my last, published, remarks on Eozoon, "Transactions of the Nova Scotian Institute of Science, Vol. IX, season 1895-96."

"The Laurentian rocks of Newfoundland are similar to those of Canada, consisting of gneiss, granite, syenite, limestone, quartzite, mica schist, etc., all of which are frequently cut by granite and other dykes. They form a large portion of the island which, as Mr. Murray remarks, 'has materially contributed to produce the remarkable geological and topographical features which it presents.' Probably it was the chopped-up appearance of the Laurentian and Huronian formation which caused him to remark that 'Newfoundland was formed of the chippings of the world.' The Laurentian of Newfoundland, so far as we know, is totally destitute of the remains of either vegetable or animal structure, and therefore must still be considered Azoic, although this term has been abandoned by some geologists in the nomenclature of Canadian rocks, owing to the discovery in the Upper Laurentian of certain forms which resemble *Stromatocerium rugosa*, one of the Protozoa of the Silurian. This peculiar mineral aggregate (?), received from Sir W. E. Logan and J. W. Dawson the name *Eozoon Canadense*. Literature enough to fill a cart has been published for and against this supposed organism, among which Dawson's "Dawn of Life" is the most interesting. With the exception of Sir J. W. Dawson, probably no one has done more work at this supposed fossil than the writer, who has prepared hundreds of microscopic sections, micro-photographs, micro- drawings, illustrative collections for the Paris, London, Philadelphia and late Chicago Expositions, and for other public and private collections ; still he could never make up his mind that *Eozoon Canadense* is of organic origin. Mr Billings late palæontologist to the Geological Survey of Canada, pronounced strongly against the organic character of Eozoon. I have frequently conversed with Dr. Selwyn, Dr. Ami, Dr. Ells, the late Mr. Vennor (who obtained the Tudor specimens), and other members of the Canadian Survey, but none of these gentlemen ever admitted that *Eozoon* is a fossil. However *Eozoon* will always remain an interesting subject for students in palæontology and mineralogy."

During part of the month of August I was again among the Bic Conglomerates trying to increase our already large fauna of fossils from the limestones of the coast rocks.

While at St. Simon, about twenty miles above Bic, I had walked many miles and worked hard to collect numerous small bivalve fossils, the muscular impressions of which were interesting various palæontologists. I had seen an Indian watching my movements during the day, but took no notice other than to wonder why he did not come close to where I was working. After supper my specimens, which I was very proud of, were packed and given into the hands of the railway station agent. While at breakfast the next morning to my surprise the stationmaster came in to tell me that my two boxes of specimens had been stolen during the night. I never saw my precious specimens again. May that wretched Indian, for I suppose it was he who stole them, be punished when he gets to his happy (?) hunting grounds!

Most of the summer of 1875, was spent collecting economic minerals for the Philadelphia Exhibition of the following year. My field of operation extended from Sarnia, Ontario, to Beauce, Quebec. With the small sum of $400.00 28 boxes of specimens were obtained, consisting of bricks, tiles, pottery, building and flagging stones, ochres, brines, salt, marbles, &c. I may mention here that through the kindness of the directors of the Grand Trunk Railway I received for about eight years, a pass which took me from "station to station." Thus the Survey was saved expending a large amount of money for travelling expenses.

Before concluding my remarks for this year 1875 it is my sad duty to record the death of my dear old chief Sir William Logan. Since his retirement he had spent a good portion of the three succeeding summers on his Eastern Township rocks. I have already stated that during the summer of 1873 Sir William was working in the Danville district and that I spent part of June with him on the Nicolet River and in other localities in that vicinity. This I believe was the last field geology he ever did. In 1874, he crossed the Atlantic to visit his friends and relatives in Wales, where he gradually grew weaker, till on the 22nd of June he died at Castle Malgwyn Llechryd, South Wales. "And now he sleeps in the quiet churchyard of Llechryd between his brother Hart and his brother-in-law

AMONG THE ROCKS.

Mr. A.L. Gower; "Peace to his memory! Honour to his name!"

The memory of Sir William will always be dear to those who have shared in his geological travels. The interesting incidents one might relate of his simple mode of life and disregard for anything other than the work he was at, would fill many pages, but I must content myself here by relating only one.

During the summer of 1864, I think it was, I spent a few days with Sir William at Point Lèvis and St. Joseph de Lèvis, opposite Quebec. Sir William was engaged tracing out some of the bands of limestone, a map of which he had already published in 1862, while I collected fossils from the conglomerates. For several days we had plodded over muddy roads, through bush and over fields; Sir William looked unusually seedy, and his broad-brimmed weather-beaten felt hat, an old coat which had seen several seasons' sunshine and rain; a pair of well worn pants, tucked in a pair of muddy boots, and a week's growth of beard, did not improve his appearance.

It was Saturday afternoon and Sir William proposed that we should go over to the island of Orleans and spend Sunday there. "All right" I said, " I will go to the hotel and settle up and you can meet me at St. Joseph's ferry." In a short time we landed at the west-end of Orleans Island, a short distance from a small hostelry kept by T. H. Lizotte. Sir William then proposed that I should interview Lizotte as to dinner and rooms for the night. I did so, and ordered dinner for two. We remained outside looking at the rocks, and in due time Lizotte came out, looked for a moment at Sir William, then beckoned me aside and said: "Dinner is ready, but do you wish your old man to dine with you, or shall I give him his dinner in the kitchen?" It is needless to say Sir William dined with me. Had I been an artist like Sir William I certainly should have portrayed the old gentleman as after dinner he sat by the fire reading a novel, now leaving off to wipe the tears from his eyes, and again to stroke a cat which sat perched on his shoulder.

Lizotte is still living. Last summer 1896, I met him in Quebec, and as is usual with him, whenever he greets me, he asked: "How is business?" I answered "I am out of business now, and Her Majesty has consented to allow me bread and butter as long as I live; but no jam." "No jam," said Lizotte, "Ah that is too bad—too bad."

Till the next summer my time was fully occupied in arranging specimens for the forthcoming exhibition. The beautiful exhibits of the supposed Laurentian fossil *Eozoon Canadense*, which attracted so much attention from geologists and palæontologists from all parts of the world, were prepared in my spare hours.

AMONG THE ROCKS.

My field work of 1876 commenced in June, at Melbourn. The Eastern Township rocks were occupying a great deal of attention and no effort on my part was spared that would throw some light on the geological horizon of these much disputed altered rocks, a thing which seemed almost impossible without the aid of fossils. I have referred to the importance of fossils while speaking of Logan and my work on the Nicolet River. It was a continuation of this work I was at when on the 13th of June, I learned by telegraph that Mr. Billings was dead. I hastened in to Montreal to attend his funeral. Not long previous to his death at the request of Mrs. Billings I left my work in the Townships and came in purposely to take him to the country for a short time, hoping the change would do him good. There was a comfortable summer hotel near the lake at Belœil Mountain, P. Q., which I chose as a quiet retreat, but we had only been there three or four days when he became nervous, and so restless that I was obliged to take him home again. This was the last time he left his house. It was with deep sorrow that I saw the last of one with whom I had been associated so intimately for over fifteen years.

Mr. Billings' funeral over, I returned to Melbourn on the St. Francis River, and as usual boarded with our old friend Mr. Gee. Poor Gee, I often wonder why Logan did not speak more frequently of this singular man, at whose house he preferred to stay, even if working half a dozen miles away. Gee prided himself on his "geological knowledge," and in his little store and grog-shop, kept numerous specimens of rocks and minerals, among which was his celebrated "fossil duck" which was a piece of water-worn limestone.

Our "grub" was none too good, but on special occasions we got, by way of a treat, an English plum-pudding. It was after being out all day, with Logan (in 1873, I think)—who never ate between breakfast and his evening dinner—that I returned very hungry, having forgotten to take some lunch, and ate very heartily of plum-pudding, the result of which was a severe attack of night-mare from which I was wakened by Sir William rushing into my room shouting "wake up Weston you have been eating too much pudding that comes of Gee feeding us too well."

Gee set apart one of the rooms of his house, which he called the "Geologic room." It was in this room Sir William often sat till the small hours in the morning, protracting his work of the day. The last night I spent with him in this room, at twelve o'clock, he was dozing over his papers, but at my request left them and went to bed; a few days after this he returned to Montreal never to work again at his beloved township rocks.

AMONG THE ROCKS.

I had only been at Melbourn a day or two, after returning from Mr. Billings' funeral, when I was recalled and requested to proceed at once to Philadelphia to take charge of the Canadian Geological Exhibit at the Centennial Exhibition. Dr. Harrington, of the survey staff, who had been in charge several weeks and whom I went to relieve remained a day to see me settled, and then left for Montreal.

My duties were to answer any questions relating to the specimens exhibited, and to keep a general supervision over our department. Many thousands of people viewed the Canadian geological exhibit every day and many amusing remarks were made about some of the specimens. The large gilded pyramid representing the amount of gold taken from the mines in British Columbia during a certain number of years, attracted much attention. It was pinched, scratched, sounded with the knuckles, and otherwise ill-treated. One old lady after rapping it with her knuckles exclaimed : " Why that ain't gold, what a sell." A man, after sampling it with his knife said, it was " wood made to look like gold to gull the public."

The fine display of amethyst quartz from Thunder Bay, Lake Superior, which was not under cover, gave me a lot of trouble. People were constantly trying to steal some of the small pieces. An elegantly dressed lady while passing

REMINISCENCES

this display took a very fine specimen and hid it under her cape. I saw the act, and although sorry, requested her to replace it.

After eight weeks constant attention to my duties, Dr. Bell, of the Geological staff, came to relieve me. I remained a day to post him up, and then returned to Montreal, not sorry to resume my quieter duties of the Survey.

I have often before alluded to the supposed Laurentian fossil *Eozoon*, but I wish to record here the interest scientific men from all parts of the world took in this supposed organism. Frequently before looking at any other objects they would say "I want to see the specimens of *Eozoon*," and so it was at the late Chicago Exhition where the finest display yet made at any exhibition was to be seen.

The winter of 1876-77 was occupied in the usual museum and office work. Many microscopic sections were made, fossils developed, etc. Mr. Billings was sadly missed by those who had been so closely associated with him, and it was long before we could reconcile ourselves to his successor, Mr. J. F. Whiteaves, late Recording Secretary to the Montreal Natural History Society, and Scientific Curator of the Museum.

A good portion of the summer of 1877 was again spent in the Eastern Townships, trying to gather material which would throw more light on the "Quebec Group," and, at the same time, to make a large collection of rocks and minerals for distribution. South Stuckley, Harvey Hill Copper Mines, Hancherchief, and other places were visited, and on the 12th of June, I accompanied Mr. Arthur Webster, of the geological staff, to Berthier, P. Q. We next went to Drummondville, the scene of my first explorations for the Survey. But the old couple who had watered Mr. Richardson's whiskey in 1863 were gone, and as my friend Webster said, " lay with the tips of their toes turned to the roots of the daisies." A good bit of new information was gained regarding the fossiliferous rocks of that locality. Kingsey and Danville were then visited, after which I left Mr. Webster, and on the 25th June I left Quebec to join Mr. Richardson at St. Jene, island of Orleans. After a short time among the rocks of that vicinity, we crossed the island to Ste. Famille, on the north side, and went carefully over the rocks to the west-end, where we again met our old friend Lizotte, who, since entertaining Sir William and myself in 1864, had built a much more pretentious hotel.

From the island of Orleans I again went to Point Lèvis and worked among the conglomerates. It was during this visit that I collected the beautiful group of *Salterella*, at St. Joseph de Lèvis, the first yet found in the rocks of the Quebec group. *Salterella* was then supposed to be a characteristic fossil of the Primordial or Lower

REMINISCENCES

Potsdam formation. Sir William had told me years ago that some day I would find *Salterella* in the lower bands of conglomerates at Lèvis, and I felt proud that his words ha come true, but regretted he had not lived to see my beautiful little group of *Salterella*.

Prof. Walcott, of the United States Geological Survey, in one of his publications, credited Dr. Selwyn, director of the Geological Survey of Canada, with the discovery of this fossil; but perhaps my name got detached from the specimen.

Leaving Point Lèvis, I next spent several days examining the conglomerates and other rocks along the south shore of the St. Lawrence, opposite the east-end of the island of Orleans The fossil fauna of these rocks are the same as those of Lèvis; but hold some forms not found at the latter place.

A short time after returning to Montreal, I again started off to investigate the rocks of the Etchemen River. Mr. Willimott, then lately of the Survey staff, accompanied me. Many new facts relating to the Quebec group were obtained. The first fossil sponge specula yet found in these rocks were discovered in the green shales of the Chaudière Falls, together with one or two small bivalve shells belonging to the *Obolella* family.

The geological structure of the Etchemen River rocks is very interesting, and as one journeys by stage or buggy up or down the river road, and puts up for the night at the various villages, he is sure to meet with incidents of Canadian life which, with the stories of the "drummer," who is almost sure to be met with at country villages— will well repay a journey through this part of Canada.

Before leaving the field for this year the following localities were visited and much work done with the hope of throwing more light on the rocks of the "Quebec group." Sept. 7th, I was at St. Garvis studying the quartzites and black slates; Sept. 13th, at Armagh; Sept. 15th, at Rivièredu Sud; Sept. 22nd, at St. Francis; Sept. 25th, at Berthier; Sept. 27th, at St. Thomas, and other places in the Province of Quebec.

It might be supposed by one not versed in geological field work, indeed, it has often been said, that our work is a series of pleasant trips, at the expense of the Government—well, let the critics try a few weeks in some of the newly settled districts of the Townships.

Field operations being over for this season, I again settled down to my work-room and museum duties. Sir William Logan was always averse to giving away geological specimens; but this year, assisted by Mr. Willimott, I commenced making up (on a small scale) collections for educational purposes; these, when distributed to various colleges, proved so important that almost from this date a constant demand has been made

REMINISCENCES

on the Survey for these educational collections; thus, what commenced on such a small scale has become an important branch of the Survey, and now, as for years past, occupies almost the whole of Mr. Willimott's time.

As with the early distribution of rocks and minerals, so the Ethnological branch of the Survey commenced on a small scale. For three or four years after my appointment to the Survey, half a dozen or so of Indian stone and other implements stood on the mantle-piece in a small room on the "fossil flat"—already referred to as being used by Sir William as a sitting and sometimes bed room. These, together with two or three casts of Indian stone pipes, and a few fragments of Indian stone pottery, etc., I had picked up near Niagara, with Sir William's permission I placed in a case in the museum; soon two cases were required, and so commenced the Ethnological collection which now numbers many thousands of specimens, hundreds of which (and some of the finest) are boxed up and stowed away for the want of room to exhibit them.

CHAPTER VII.

THE SOUTH SHORE OF THE ST. LAWRENCE—SOME GOOD FISHING—CAMPING BY THE RIVER—ASCENT OF THE SHICKSHOCKS—A POACHING EXPEDITION : WE CATCH A FINE SALMON—THE LONELY MOUNT ALBERT—PHOTOGRAPHS AND FOSSILS—COAST ROCKS BETWEEN DIGBY AND YARMOUTH, N.S.—"THE OVENS"—FISHING AT SPINDLE COVE—LOST IN THE FOG.

ON the 18th of June, 1878, with camp equipment and two months' provisions for myself and one man, I left Montreal for Gaspé. My office instructions were to examine the coast rocks on the south shore of the St Lawrence River between Cape Rosier and Metis, especially with regard to the fossil fauna.

At Gaspé I was fortunate enough to secure the services of the Indian John Basque, who had in 1843 accompanied Sir William Logan over the same ground. John (he is dead now) was a tall, straight, well-built man, and handsome for an Indian. Besides speaking English and French, he spoke the language of several different tribes of Indians, but could neither read nor write his own name, things he very much wished to do and which I tried hard to teach him, but it was no use, and he gave it up for a bad job.

We made our first camp a short distance from Cape Rosier lighthouse, and after a good supper of fried ham, sea-biscuit and tea, John collected

REMINISCENCES

from a neighboring bush sufficient spruce-boughs to cover the floor of our tent. On this our blankets were spread, but John seldom came inside, preferring to make a wigwam for himself.

The summer nights of the Gulf shore of the St. Lawrence are always more or less chilly, but drift wood is plentiful in most places, and one can always, without the use of the axe, keep up a rousing fire.

I found John a pleasant companion, and I sat at our camp fire and watched the dark clouds spread over the distant mountains.

All was solitary and still excepting the everlasting splash of the waves on the pebbly beach before us. It was long after darkness had obscured every object within a few yards of our camp fire before I tumbled into my blankets. When I awoke in the morning I found my man busy preparing our breakfast. John, who never lost an opportunity of making new friends, had already been to the lighthouse and brought back some fish. After breakfast we closed up our tent, and with a little lunch in our collecting basket, a gun, hammers and note-book, started on this the first part of our geological tramp of over one hundred and eighty miles along the base, and frequently over the tops of the great cliffs of the Gulf shore.

My examination of these exposures commenced at the base of the cliffs almost under Cape Rosier lighthouse. Here the strata consists of conglomerates, grey limestone bands, black bituminous limestone, greenish and other coloured shales. No fossils had hitherto been found by which the geological horizon could definitely be determined, but they were supposed by Logan to belong to the Hudson River formation. It was my good fortune, however, on this the first day's work to find, almost in the shadow of Cape Rosier lighthouse beautiful compound Graptolites which evidently belong to the Lèvis zone—which according to the latest nomenclature of the Geological Survey of Canada is Upper Cambrian.

The sun was setting when we returned to camp. I think John thought we had done a little too much work for one day, but I assured him that I should not work so hard every day and that he would not have to work on Sundays, as with Logan, excepting to cook—a thing an Indian is always ready to do.

It was long after dark when I crept into my blankets, but before doing so I had put some dry plates in the dark boxes of my camera which I had brought with me—hoping to get some good geological views of the coast.

I may mention here that I did get many fine views, some of which have become historical, having been published in the reports of the Survey. I may also mention here that all the negatives taken by the Survey staff are carefully preserved. They have lately been re-arranged, catalogued, numbered and placed where any one of them can be found at once, by Mr. Percy Selwyn, who is now private Secretary to the present director of the Survey, Dr. G. M. Dawson.

It would extend these notes too much to recount all the adventures John and I met with while journeying along this great stretch of sea coast. We shifted camp frequently, generally every eight or ten miles. By so doing we were able to walk down the coast to where we left off our examination, and up the shore to where we would camp next. Our mode of shifting camp equipment, etc., was by boat or cart, whichever we could get most conveniently.

During our journey by boat I frequently had some good fishing. On one occasion while sitting at the stern of the boat. I noticed fish jumping—at flies, I suppose; but I did not see any. Having a trolling line with hooks and spoon, silvered on one side and red on the other, I was not long in getting it ready. The two boatmen nudged each other and winked, while John gave a sarcastic smile, but I let go my troll, and before the men had pulled half a dozen strokes I hauled in a lovely mackerel, and before we completed our ten miles' journey, twelve fine mackerel and two cod fish lay at the bottom of our boat. The men were greatly surprised and so was I. Fishing for mackerel and cod with a trolling spoon was a new feature on the St. Lawrence.

Landing at Gritton Cove, about ten miles above Cape Rosier we pitched our tent at the base of huge black cliffs composed of shale interstratified with other rocks. The escarpment represented here is probably 1,000 feet thick. Many of the beds are highly fossiliferous, some being very prolific in Graptolites, which have a decided Hudson River aspect.

In the late publications of the Geological Survey of Canada, the formation here and for about eighty miles of the coast to the Marsouin River is Combeo Silurian i.e., Trenton, Utica, &c. The late changes made in the geological horizon of the coast rocks are chiefly due to fossil evidence collected by the officers of the Survey, since the time of the late Sir William Logan. All the fossils collected on the journey of which I write, and late journeys of Dr. Ells, Dr. Selwyn and myself, have been studied by Prof. Chas. Lapworth of Birmingham, England, who I believe is the best English authority on the Graptoliæ fauna.

The coast country between Cape Rosier and Cape Chat, a distance of about 130 miles, is mountainous. The hills and cliffs come close on to the shore leaving for many miles no space for settlement, excepting at the mouth of rivers where one finds fishing stations of more or less pretensions. We always received a warm welcome from these Gulf shore people who seldom see any other than their own class.

We made many camps between Griffon Cove and Cape Chatte, and gained much information regarding the fossil fauna.

At Ste. Anne, ten miles below Cape Chatte I found my old friend Mr. James Richardson of our Survey staff camped near the foot of the river. Mr. Ord also of the Survey staff, Mr. John Richardson and one Indian had just returned from the foot of the Shickshocks mountains, a distance up the river about 30 miles not taking in the many crooks of the stream.

As it was my intention to ascend one of the highest peaks of the Shickshocks, and as it required two experienced men to pole a canoe up this river I engaged the Indian Joe, who had just returned with Mr. Richardson. But Joe flatly refused to accompany me till he had been drunk, at least one day. John, my Indian, assured me Joe would turn up all right according to promise, said he "If Joe says a day, he means a day, and no longer." It was impossible to get another man who knew the river like Joe, for he had been tripping on that stream for years, so I had to submit to Joe's terms. A certain hour was fixed for his return to camp, and to my surprise—but not to John's, Joe turned up all right, and when asked how he felt after being

AMONG THE ROCKS.

drunk a day and night, said, "Bully." I preferred a long narrow boat, much used on this river, to the canoe Richardson had used. This I obtained from the settlement at the foot of the stream, and we were soon ready to start. We took our blankets and three days' provisions, and of course a gun and fishing tackle. We lunched off a magnificent trout Joe stole from a net. It was the most delicious fish I ever ate.

The Ste. Anne River is, or was, at the time I ascended it, one of the finest salmon rivers in Canada. In many parts the current is swift and requires dexterous poleing and in places portaging, but in places one passes over still pools where the paddle can be used. As we glided over these pools and looked down into the clear water we could see many salmon quietly floating or resting on the bed of the river.

It was dusk when we hauled up for the night, ten miles from our starting point.

After supper John and Joe collected a lot of birch bark which was soon made into a flambeau. Then Joe fished out from some hiding place, a salmon spear—which if found on him at the settlement would have cost him lots of trouble. When the night was as dark as it would be, we stole forth and silently glided down stream. Then I realized that for the first time I was on a poaching expedition—for the fishing rights of the Ste. Anne were very strict, and I had no permit. But almost before I had time to reflect on our evil ways, a twenty pound salmon lay at the bottom of our boat. I forbade the men to take any more fish than we needed for a change of diet.

I shall never forget this my first poaching experience, the stillness of the night, the brilliant light of the flambeau reflected on the eager faces of the two Indians, made up a scene well worthy of the artist's brush.

The following morning we resumed our journey. Towards noon the sun was very hot and the weather most oppressive. John had been standing at the bow of the boat fully an hour assisting Joe to push through one of the most difficult parts of the stream, to navigate. Suddenly he let go his pole and sank to the bottom of the boat. Instantly our boat swung round and shot down the swift-current, but in a short time Joe ran it ashore and we hauled John out to a shady spot and laid him on his back, and for the next two hours I thought we were going to have a dead Indian to take back. We came to the conclusion that John was suffering from sun-stroke. Remembering I had a bottle of "Pain Killer" in my fishing basket I made up a strong dose, but it was a pretty big job to get John to take the white man's medicine. He did however, at last, and in the cool of the evening we resumed our journey.

AMONG THE ROCKS.

The third evening after leaving our camp, near the mouth of the Ste. Anne, we reached the foot of the Shickshocks and camped for the night in a trapper's deserted wigwam. At six o'clock next morning we breakfasted on the remains of our twenty pound salmon and then prepared for the ascent of Mount Albert. John carried my camera and other traps wrapped in blankets, while Joe carried two days' provisions and camp utensils. I carried—well—myself, a canvas bag and small hammer, which I found quite enough. I could not help wondering that while I stumbled sometimes head first over fallen trees or slipped down moss covered rocks, my two Indians glided through all the difficulties which beset us without a slip or stumble.

In about two hours we reached the summit of Mount Albert 3,768 feet above the sea—a great dreary table-land with a few stunted spruce trees, and strewn with weathered rein deer (?) antlers. We chanced to reach the summit of this mountain within a short distance of where Logan and Murray in 1844, erected a flag-staff and unfurled the Union Jack. The pole was still lying at the foot of the mound of stones which had supported it, but it was broken and weather worn so John descended the mountain till he found another good stick ; we then rebuilt the mound, hoisted our pole and in place of a Union Jack substituted a red cotton handkerchief.

REMINISCENCES

The magnificent panorama presented from the summit of Mount Albert is grand indeed. The eye ranges over hills and valleys for a hundred miles or more. The river we have ascended looks almost like a brook winding through mountains and forest till it joins the St. Lawrence, the waters of which are dotted with ships and fishing boats.

Although we can see some life in the distance, I shall never forget the sense of loneliness I felt. Even the Indians while we sat round our camp fire spoke in whispers as though evil spirits wandered round.

Crossing the table-land to the south side of the mountain, we look down over masses of serpentine and other rocks, into a great valley in which a small lake nestles. I set up my camera hoping to get a view of this wonderful scene; but the black flies are so numerous that they partly cover my lens and I see by my focussing glass that a number have got inside the camera. I look up to speak to Joe and find he is clasping a small crucifix which hangs from his neck. He is actually trembling with fright. "Joe what on earth is the matter with you?" I say; Pointing to the lake, he says in a whisper "Look— see the devil is taking a swim." I look and find a strong current of wind is passing through the valley, which accounts for the heavy ripples on the water. I asked Joe if he had any name for that lake and he whispered yes, "that devil lake." All I could say to this poor fellow would not alter his opinion regarding the devil and his ablutions.

Granites enter largely into the formation of the Shickshocks, but hornblende, quartzites, epidotic and other rocks are largely represented, together with massive beds of serpentines, which are frequently beautifully stratified. Pieces of chromic iron ore, about the size of one's fist, are scattered over the table-land of Mount Albert, but I could not find the bed from which they came.

We only made one night camp during our return journey—a journey full of pleasant remembrances, and of much geological interest.

Arriving at our old camp near the mouth of the river, I found Mr. Richardson waiting me. Poor Mr. Richardson, the veteran explorer for the Geological Survey of Canada almost from its commencement—a man who had worked his way from a poor farmer to that of the most trusted field geologist then on Logan's staff, had received an intimation that the present director of the Survey was about to apply to the government for his superannuation. This, to a man like Richardson, was a blow indeed, for outside his geological work he had little or no pleasure. He remained with me several days, during which time we visited two or three interesting fossil localities on the coast, one of these was about three miles above the Ste. Anne. Here we obtained large slabs of black slate covered with the well known Lèvis fossil *Phillograptus typus* (Hall), and a number of other species. After a few days Mr. Richardson left me and returned to Montreal. John and I continued our journey along the coast, camping as usual about every ten miles. Many new fossil localities were discovered, and forms found which were new to Canada. Duplicates of all the fossils collected were sent to Prof. Charles Lapworth, of Birmingham, England, for identification and description.

During this long sea coast experience many photographs were taken, prints of which can be seen with the large collection of photographs in the Geological Survey rooms. One of these views shows the " Pillar Sandstones " on the coast, eight and one half miles east of Ste. Anne River. I focussed my view, then left John to expose the plate, so that I might appear in this picture with a little Indian dog who came to our camp one day, whence we did not know. He was first seen sitting on his hind legs, as if asking to be taken in. John at once adopted and christened him Wap-e-cat, which he said was the Indian for "white paw." This picture is reproduced in the Geological Report for 1880-81-82, but by some mistake is credited to Dr. Ellis, 1883.

I am loath to close this brief account of a journey during which every day brought to life some new geological information, besides giving one an insight into the life and habits of the poor fishermen, many of whom labor year after year and never see anything other than their immediate surroundings.

We reached Little Metis Sept. 21st, where we struck camp for the last time. I paid off my Indian, John, saw his money safely sewn up inside his vest, and started him off to his wigwam at Gaspé.

The winter of 1878-79 was spent in the usual museum and and work-room duties and in some preparations for the removal of the Survey to Ottawa.

My first exploration journey of 1879 was to Roxton Falls, P. Q., to examine the limestone and shales associated with the copper deposits. June 5th, I left for Nova Scotia to examine the shore rocks of St. Mary's Bay. The journey from St. John, N. B., to Annapolis by steamboat—45 miles across the Bay of Fundy is, in calm weather, a pleasant one, but this time it was most unpleasant. From the time we left St. John, till we arrived at Digby Gut—the entrance to Annapolis Basin—the waves frequently swept over our boat. All passengers were below deck, and most of them sick. One old body was sick unto death, and if ever she reached dry land again, &c., &c. I am a good sailor and did justice to my 75 cent dinner.

The occurrence of fossil remains in the gold

REMINISCENCES

bearing rocks of Nova Scotia, by which the geological horizon could be determined, would be an exceedingly important discovery. From time to time forms which were thought to be of organic structure have been found, as the following remarks by the writer, taken from " Transactions of the Nova Scotian Institute of Science," Series 2, Vol. 1, will prove. But in all cases a microscopic examination of these forms has proved them to be concretionary bodies—

"Many times between the years 1860-70 the late Sir Wm. E. Logan, and subsequently Dr. Selwyn, called my attention to certain concretionary forms found in the goldbearing rocks of Nova Scotia. Some of these seemed to be organic, and I was requested to make and examine microscopic sections of them. In treating several of these with acid, they proved to be composed chiefly of dolomite, with a large proportion of siliceous matter, and generally a little iron pyrites, which formed a nucleus.

" In 1890 a number of similar forms were found by Mr. Willis, in the rocks of the Northup Gold mines, Rawdon, Nova Scotia. They were handed to Professor Hind, who supposed them to be fossils, and assigned them to Lower Silurian age. Wishing the "judgment of a specialist," he gave them to Professor Kennedy, of King's College, who confirmed Professor Hind's opinion and pronounced the "fossils" to be *Stromatopora*. Mr. Fletcher and Mr. Faribault, of the Geological Survey, while in the vicinity where this discovery was made, visited the mines, and brought away a number of these so-called fossils. They were given to me by Dr. Selwyn, director of the Dominion Geological Survey, for microscopic examination ; and I regret to say, the result is precisely the same as for those examined thirty years ago.

AMONG THE ROCKS.

"They appear to be composed of dolomite, and, when dissolved in hydrochloric acid, leave a good percentage of insoluble matter, probably felspar and silica. It is likely that they were spheroidal or ovoidal in form before being flattened by the pressure of overlying beds. One of the specimens before me is a piece of greenish-grey laminated mica-schist five inches long and one inch thick. Inclosed in this are four of these concretionary forms broken through the centre, each measuring one inch in length and half an inch in breadth. Two of these are connected with each other by a thin strip of the material of which they are composed.

"In broken sections some of these bodies show slight concentric layers which in microscopic sections are not seen. Not a trace of organic structure was found.

"I quite agree with Professors Hind and Kennedy as to the importance of finding fossils in the auriferous rocks of Nova Scotia, and trust they may be more fortunate than I.

"It is well known that concretions occur in all rock formations. One or two instances will be worth recording to show how careful one should be in referring any forms of a concretionary nature to organic structure (Read Nov. 9th, 1891.)"

It was chiefly to try and discover fossils in these auriferous rocks that the present journey was made.

Gold-bearing deposits occupy a great portion of the Atlantic coast of Nova Scotia. The rocks consist of greenish and grey slates, quartzites and other rocks, which, if as stated in the Geological Survey Reports, belong to the Primordial Silurian and Cambrian formations—should contain fossils, but at present nothing with decided organic structure has been found. The so-called *Eophyton* from the " Ovens " gold-bearing slates is in my opinion not of organic origin.

Many miles of coast rocks between Digby and Yarmouth were examined, especially those of Weymouth, Church Point, Montegan, Salmon River, Bear Cove, Cape St. Mary, Cranberry Head, &c.

Having a good camera and dry plates with me I was able to obtain excellent photographs showing the stratification, folding, &c. of these rocks. The Nictaux iron ore deposits were also visited. Here the Hematite ore-beds contain well defined fossils which belong to the Devonian formation.

I travelled along the coast road by stage or buggy. The journey through the Annapolis valley is a most pleasant one. It is the great apple raising country of Nova Scotia. Continuing along the coast and shore road I reached Yarmouth, where after a few days' work among the rocks of that vicinity I left for Lunenburg—the oldest British town in the Province of Nova Scotia and on the 9th of July, crossed the harbor to the " Ovens," already mentioned with regard to the so called *Eophyton*, (Torell), *eos*, dawn ; *phyton* a plant. From the harbor of Lunenberg are seen on the west side a range of cliffs containing a number of even cavities, the regularity of which—whether made by human agencies or by the action of the sea—no doubt suggested the name "Ovens."

These rocks were known to contain gold many years ago, and considerable excitment has at various times prevailed, when a specimen of quartz or slate had been found which held a good sample of the precious metal.

The only mining going on at the time of my visit (1879), was done by a few of the fishermen of Spindle Cove, on the "Ovens" shore, who with a rude cradle, shovel, or other simple implements gathered from the sea sand of the shore, gold dust to the amount of from fifty cents to a dollar a day. But my visit to the Ovens was not for gold; but for fossils, which after a most diligent search I failed to find, nor could I find another specimen of the so-called *Eophyton* marks, which could readily be made by a bit of rock passing over a soft substance such as these slates were once.

At the time of my visit to the "Ovens" there were only five or six families living near the shore; they were fishermen. 1 found board and lodging with a German family at Spindle Cove. Among my accounts of this journey I find the item: Four days' board and lodging at "Ovens" $2.00, but besides this I paid $2.50 for boating and assisting with my camera. This small sum of $4.50 to these poor people meant much. When shown to my room the first night

REMINISCENCES

of my stay with these fisherfolk I found there were two feather beds only. Thinking they had forgotten the sheets and blankets, I found my way down stairs where already they were at their family devotions. Waiting outside the room till prayer was over, I then requested the man to go up to my room where I explained that there was no covering to the bed. He looked puzzled and called his wife up, when I made known what I wanted she looked surprised, but a happy thought seemed to strike her, and she went away with a smile on her face, soon she returned with a quilt, but at the same time thought I would find it more comfortable to sleep between the feather beds.

I have spoken of catching mackerel with a hook and spoon, but that fishing was tame to the sport I had at Spindle Cove. Shoals of herring—the well known Yarmouth bloater—came into Lunenburg harbor, followed by the horse mackerel even into the fish traps, or seines. It is a powerful fish and much dreaded by the fishermen who have herring nets or seines set, and who lose no time in getting them out of the way by harpooning.

During my stay at Spindle Cove I assisted in harpooning two of these creatures which were at last killed by severing the caudal vertebræ with an axe. One of these measured five feet long and was over two feet in diameter. I had it

AMONG THE ROCKS.

towed to shore and then photographed it. In appearance these huge fish are just the same shape and color as our mackerel; but the flesh is coarse like that of the seal.

On the 12th of July, I crossed the harbour again (four miles) with the people I had been staying with. They, at least the man and his two daughters, were taking fish to the Lunenburg market. There was a light fog when we started which soon increased till we could not see a yard before us. At last the man and his daughters ceased rowing, confessed they were lost and did not know where they were. Fortunately the man knew the compass bearing from his shore. I looked at my compass and found we were headed almost in the opposite direction. This however, these folks could not believe, but at last decided to be guided. After pulling an hour or so through the dense fog we ran right against the wharf at Lunenburg. I wished my companions of the fog good-bye, took steamer and was soon at Halifax, were I spent a short time with my friend Rev. Dr. Honeyman, curator of the Provincial Museum, who had been in 1876 my companion for eight weeks at the Centennial Exhibition, Philadelphia.

While on this journey I had received official instructions to visit on my way back, the Joggins shore and get photographs of various cliffs, to illustrate a paper Sir William Dawson was

REMINISCENCES

writing for the Royal Society. I was fortunate in getting what Sir William wanted, and these were published in the Royal Society Reports. I reached Montreal on the 20th of July.

On the 11th of August I started out again to the Eastern Townships, visiting West Farnham, Bedford, St. Dominique, Richmond, St. Pie and again to West Farnham. October 19th, I accompanied Mr. Richardson to St. Armond and Grand Ling. This was the last official journey Mr. Richardson made. He was superannuated—greatly against his will—shortly afterwards, This completed my field work for 1879.

The winter of 1879-80 was occupied in my usual microscopic and museum work. A great number of microscopic sections of rocks and fossils and preliminary examinations were made. In the museum many new fossils were labelled and arranged in the cases. In some of the work I was assisted by Mr. Willimot.

CHAPTER VIII.

BAIE DE CHALEURS AND ADJACENT DISTRICTS—A FOSSIL FISH AND A PIOUS CRITIC—A LIST OF TITHES—REMOVAL OF THE HEADQUARTERS OF THE SURVEY FROM MONTREAL TO OTTAWA—THE JOGGINS COAST, N. S.—CAMPBELLTON AND THE RESTIGOUCHE RIVER—A TRIP TO SAULT STE MARIE—THE CYPRESS HILL AND THE NORTH-WEST—A THREE MONTHS GEOLOGICAL EXPLORING TOUR.

PART of the summer of 1880, was spent in the examination of the shore, rocks of Baie des Chaleurs, Cascapedia and Matapedia Rivers, Campbellton and other localities in that district.

The shore rocks of the Restigouche River near the back of Campbellton Railway Station and those on the opposite shore are highly interesting, containing as they do, a remarkable fauna and flora of fish and plant remains of Devonian age. Many fine fossils were collected from this locality by myself and later by Mr. Foord.

On the 11th of June, I crossed from Dalhousie to the north shore of Scaumenac Bay where I got board and lodging at a farm house, the occupants of which were an old couple and a grown up son and daughter.

My field of research here was along the shore towards Pt. Maquasha, Baie des Chaleurs, but chiefly on the shore of Scaumenac Bay. Here high cliffs of light yellowish grey sandstone occupy

REMINISCENCES

the shore. The rocks are of Devonian age and underlie the lower Carboniferous conglomerates of the north shore of the Restigouche River. It was in these Devonian cliffs of Scaumenac Bay that Mr. R. W. Ells of the Geological Survey in 1879, discovered that curious crustacean *Pterichthys*; or fish belonging to the Ganoids—whichever it is. This discovery led to my present researches and during the next summer to those of Mr. Foord, also of the Geological Survey staff.

Some of the beds of these great sandstone cliffs contain nodules, or concretionary forms. The following description from my notes on concretionary forms published in Transactions of the Nova Scotian Institution of Science will suffice.

"The Upper Devonian fish and plant-bearing beds of Scaumenac Bay, New Brunswick, are prolific in fossiliferous concretions, which are composed of calcareo-arenaceous rock, and take various forms according to the shape of the nucleus, which, when a fish, is often so well preserved that every bone can be seen. One of these concretions obtained by A. H. Foord measures over twenty-one inches in length, and contains the skeleton of a fish almost as long. It is *Chirolepis Canadensis* (Whiteaves). In other concretions from this locality the writer and Mr. A. H. Foord found : *Glyptolepis microlepidotus* (Agassiz), *Phaneropleuron curtum*, *Pterichthys Canadensis* (Whiteaves), *Eusthenopteron Foordi*, etc."

These concretions are scattered along the shore, being washed up by the waves of the bay. It was in one of these that I discovered the first fossil fish known to the Survey, from these

AMONG THE ROCKS.

rocks This find is always associated with the old couple with whom I boarded. When in the barn carefully chiselling off a portion of rock which concealed part of my fossil fish, the old lady came in, and after watching me for a short time said : " Well now, it do look like a mackerel, but if God made stone fishes it was for some wise purpose that we poor mortals can't understand and ought not to meddle with, and Sir, it would be much better for you to leave them where the Lord placed them." I tried to explain that a few million of years ago this fish swam in the sea, then died and was buried in the sand and mud at the bottom, then the sand and mud became stone. But the few millions of years seemed to frighten the old lady and she left me. Then the old man came to interview me on the subject, and chided me for even thinking of a few millions of years ago, said he "Do you dispute the words of the Holy Bible, etc., etc." That evening I was requested to attend service before retiring for the night. The old man read a chapter from the Bible explaining the same as he went on, then the family sang a long hymn, drawling out each word till I dozed off to sleep. But at last singing was over, and then the old man delivered what some church people would call a most powerful prayer, full of beautiful thoughts, which ended with "Oh Lord bless the stranger within our gates, and keep him from vain babbling; and turn his mind from things of the past, to his future salvation. Amen." No doubt the few millions of years was in the old man's thoughts when he spoke of vain babbling.

The following day I collected from these same fossil fish-beds, fossil plants, among which were specimens of an old fashioned fern—which Sir William Dawson has since named *Archaeopteris Jacksoni*. This specimen puzzled the old man of my boarding house, but no further allusion was made to vain babbling in the old man's prayer that evening.

Leaving Scaumenac Bay July 1st, I proceeded to New Richmond, where a few days were spent on the rocks of that vicinity, a journey was then made to Causapscal railway station, on the Intercolonial Railway, and on the Campbellton road, near the Matapedia River. Here I obtained lodging at the trackman's cottage, where I fed on fat pork, brown bread and potatoes, while H. R. H. Princess Louise—two or three hundred paces away—lived on the best of the land, and while I fished with my hammer among the rocks for fossils, she fished in the river close by, for salmon; and got them, too. Several members of the Royal family have been the guests of Sir Donald Smith, who has a summer house or fishing station on the banks of the Patapedia River, a lovely spot close to the Causapscal Ry. station, which, since being patronized by royalty, has

AMONG THE ROCKS.

become quite a noted spot on our long railway. The platform of this station is the lounging place for Indians while waiting to be engaged by sportsmen who go up the Matapedia River, either to fish or shoot.

Having completed my examination of the rocks at the Devil's Elbow, a turn on the river two miles or so below the station, with a canoe, two Indians and a few day's provisions, we left for a journey up a portion of the Matapedia River. The rocks of this river belong to the Gaspé series, and are almost destitute of fossils. One who has camped on this beautiful river will never forget the charming scenery, and if he is fortunate enough to "hook" a salmon or one of the large trout found in the rivers of this vicinity, and to have his Indian canoe-men cook it in their fashion, he will long remember his camp on the Matapedia River.

Arriving at Campbellton on Saturday evening, I concluded to stay there until Monday. Judge ———— who was on his circuit, and staying at the same house, invited me on Sunday morning to accompany him to service at a small Catholic church on the north side of the river. We engaged an Indian to paddle us over in his canoe. The little church was already crowded with people of various nationalities, with a good sprinkling of Indians.

After the sermon—in which the priest exhorted

his flock to attend to their religious duties more diligently, on pain of excommunication—he produced a roll of paper in which were the names of those who had and had not paid their tithes: John Brown, four cords of wood; Patrick O'Farity, ten pecks of potatoes; Peter Basque, fifteen pecks of beans; John Gabriel, one side of pork; Francis Cye, one ton of hay; Narcisse Cromk, NOTHING, and the priest looked round with fire in his eye, but he could not spot poor Narcisse, and so the list went on.

My wanderings during the remainder of the field season of 1880 covered many hundreds of miles, and new geological facts were obtained from the rocks at Father Point, Rimouski, Negette, Bic, and many other localities on the St. Lawrence shore and in the Eastern Townships.

The most important event connected with the Survey this year, 1880, was the removal to Ottawa.

The director's summary report for 1881 states that: "The total number and weight of packages forwarded from Montreal between the month of November, 1880, and May, 1881, was 1,729 boxes; 101 barrels; 162 miscellaneous packages—gross weight, 282,585 lbs.

The work of packing all the type fossils from the cases, specimens from the drawers, wall cases and the large wall specimens, was done by myself and those under my direction. In this

AMONG THE ROCKS.

work I was assisted by Mr. Broadbent and Mr. Holmes. It gives me great pleasure to record here the valuable services rendered in this work by Mr. Broadbent, a young Englishman, who had just come to this country to seek his fortune, and who, with the Director's permission, I engaged to assist in the removal of the fossil department of the Survey. Mr. Broadbent being a good writer, I appointed him the task of cataloguing the species and formation of every type fossil taken from the cases—a work he accomplished most satisfactorily, indeed, his diligence in the duties assigned him decided me—with the Director's permission, to take him to Ottawa to assist in reorganizing the collection. After assisting me for several months, Mr. Broadbent was removed to the mineral department, and was soon after appointed a member of the staff. He now holds the position of museum assistant in the mineral department.

The work of replacing all the type specimens in the cases, and the thousands of specimens kept in drawers under the cases required much time and patience. Mr. J. B. Tyrrell. so well known now as field geologist and explorer in the distant lands of the North-west Territory, assisted me in part of this work.

After spending most of the summer months since 1863, in travelling and field work, it was not pleasant to spend the whole of the summer,

REMINISCENCES

excepting five days in museum work. The five off days were, at the Director's request, spent in accompanying Mr. Walter Billings, Mr. Ami and Mr. Souter to Paquett's Rapids, on the Ottawa river. It was dark when we arrived, with our camp outfit and a few days' provisions at our destination on the 5th of September. My three companions were students in Palæontology, and so eager to obtain some of the fine Black River fossils which occur in the Paquette's Rapids rocks, that instead of pitching tent and getting off to bed, they started off with a supply of matches to hunt fossils. To an old collector like myself this was a strange procedure. When they returned I was rolled in my blanket and sleeping. Before my companions were up the following morning I had collected one of the finest specimens of *Stromatocerium rugosum* now in the geological museum.

Mr. Ami, now Dr. Ami, M.A., etc., etc., is assistant palæontologist to the Survey, and is one of the bright scientists of the Survey staff. I had a jolly time, and returned with renewed energy to deal with the *Climactichnites* tracks of which I spoke in connection with Lord Dufferin's visit to the Survey museum.

Unfortunately the director who superintended the taking down of this large specimen had not calculated its weight, and in lowering it from the wall, when at an angle of about 45° it fell with a crash, shattering the plaster casts round the specimen into fragments and breaking some of the sandstone. The director got a black eye—which he afterwards told his friends was not a disreputable one, and the two men assisting got hurt. It is a wonder it did not go through the floor to the next flat. I was then requested to take it in hand. The fragments were boxed up and shipped to Ottawa, and after three weeks' work, assisted by Mr. Broadbent, a carpenter and a man who made new casts to fill up the frame, we succeeded in getting it placed against the wall once more. We made the best it was possible to make of it, but the specimen, which was one of Sir William's pet fossil slabs, is far from looking as well as when in his bed-room in the Montreal museum.

Before and while the removal of the Survey was in progress, the families of the members of the staff were removed to Ottawa at the Government's expense.

After seeing my family settled in the capital, I returned to Montreal, and boarded, strange to say, in the "Logan House," which had been rented by the administrator of the Logan estate, Mr. Grant, to my friends the Austins. The room assigned me was the one formerly used by Sir William as library and sitting room, a delightful room opening by French windows on to the spacious garden. I spent many happy evenings in this old homestead where there were still

REMINISCENCES

things left to recall the memory of our dear old chief, and the time when he was ever ready to welcome the humblest members of his staff, and to listen patiently to their troubles. I recall the time, one Sunday in 1865, when oppressed and despondent, I went to tell him I thought of returning to England. "Fiddle-de-dee" said Sir William. He then like a father reasoned with me and pointed to a brighter side of my troubles. When I left his house it was with renewed energy to continue my duties at the Survey. Such was the kind influence ever extended to those with whom Sir William had to deal.

Many other incidents worth recording occurred during the year of our removal to the capital, but I must hasten on with my travels.

By the spring of 1882 the geological museum was in fair order and my field work was resumed early in June, by a journey to the Joggins coast, N.S., with a view to increase, by fossils, our knowledge of the Carboniferous formation.

To the student who wishes to study the geology and palæontology of the Carboniferous formation there is probably no better opportunity afforded in the world than the Joggins coast. Here he sees a magnificent range of cliffs extending for miles along a shore washed by the waves of the Bay of Fundy.

Occasionally in the great red sandstone cliffs one sees portions of erect trees, or rather the casts of portions of large exogen trees, the woody structure of which decayed and disappeared leaving the bark standing, to be filled with the sand of the sea or lakes near which they grew. While in a decayed form the base of some of these trees were penetrated by small reptiles who no doubt took shelter there for safety. These little reptiles represent the first vertebrate animals of this world. Some of the sandstones and shales contain the remains of beautiful ferns and tropical plants, on the leaves of which we find small shells, representatives of the Mollusk family. Other beds are made up almost entirely of small bivalve shells, which Sir William Dawson called *Naiadites carbonaria*.

In the geological museum can be found a fair representation of the Carboniferous fossil fauna and flora of Canada; but—as my last discovery in the Joggins rocks proves—there is still much information to be gained with regard to the life of the Carboniferous period.

The splendid geological work accomplished in the Carboniferous formation of Novia Scotia by Logan and Dawson in the early days of the Survey, and later by other members of the staff, is a most important feature in the geology of Canada, for on the correct interpretation of the Carboniferous strata often depends vast sums of money.

Having secured fine sections of the fossil trees, *Sigillaria*, so named from the seal-like scales on the bark,—and specimens among which were scales of fishes, ferns, and other fossil plants—work in which I was assisted by Mr. James Devine, a coal miner, I spent a few hours visiting the mines, which are situated half a mile or so from the coast.

In the superintendent's house I was surprised to see several mounted moose heads *(Cerius alres)* which my friend the superintendent said he shot at his hunting grounds a few miles from Joggins.

At this time there was only one main road in the village and another, the coast road, which ran near the brow of the cliff. On these roads are dotted the homes of the miners. The coal is conveyed from the mines by tram-ways and dumped down a chute to schooners and other vessels

Fifty years ago few people excepting those engaged in mining, or in the manufacture of grindstones were to be seen in the village, and along the Joggins shore. All communication being by stage, buggy, or boat.

Having packed and shipped my specimens I again mounted the rambling old stage and after a pleasant ride through a verdant country over which refreshing sea breezes are wafted from the Bay of Fundy, I arrived at Maccan Station, and on June 26th, again landed at Campbellton where a few days were spent collecting from the

AMONG THE ROCKS.

Devonian rocks on the banks of the Restigouche River.

Among the interesting specimens secured this time were good examples of the fossil-plant *Psilophyton princeps*, (Dawson) Miller says it is the oldest known plant in America, and is supposed to have grown in a marsh (*psilon*, smooth; *phyton*, stem.)

The agglomerates of this vicinity are rich in fish remains, and the conglomerates with their various bright colored pebbles of jasper, sandstone, quartz, agates, and other rocks, enhance the beauty of this lovely spot where not so very many years ago the red man alone pitched his wigwam, built his birch bark canoe, and speared salmon as he glided down the often swift waters of the Restigouche to one of his wigwams, where now stands the little church of which we have read in previous pages of these notes. Now the whistle of the locomotive echoes among the hills and through the valleys, the white man plies his axe, Royalty dips its fishing-line in the waters, and the geologist pokes his nose—I mean his hammer into the rocks.

A few days after my return to Ottawa I received instructions to proceed to Sault Ste. Marie, to examine and collect typical specimens of the rocks of that locality, and along the coast of Georgian Bay and part of the north coast of Lake Huron.

REMINISCENCES

It was then, and is now, an important question as to the exact geological horizon of the red sandstones, so largely displayed in the construction of the Soo Locks. It is supposed that they belong to the Potsdam formation, but the want of fossils to confirm this question still remains, as my researches revealed nothing which could definitely be pronounced organic. It is recorded that one fossil of a Potsdam type was found in some part of these sandstones; but as no one knows when, where, or who collected it, I think we may consider it doubtful. Still I think with Logan that these sandstones are Potsdam, and that fossils will be found to confirm this fact.

My short stay at the Soo was full of pleasant incidents most of which were due to Mr. Cousins, a prominent Civil Engineer of that place. But our pleasant moonlight boating parties—like all other good things, came too soon to an end.

Leaving Sault Ste. Marie July 22nd, with two half-breed Indians and a boat, a journey was made down that beautiful stream called Garden River—the highway for ships passing through the Soo Locks to Lake Superior. Returning to Sault Ste Marie a few more days were spent in the examination of the rocks of that vicinity, after which, providing myself with a horse and cart and an Indian guide, camp outfit and provisions, a journey was then made by road to

AMONG THE ROCKS.

Echo Lake. All rock exposures were noted and typical specimens collected.

Arriving at Echo Lake, I was cordially received by those in charge of the copper mines property, which is situated on the north side of the lake, two or three miles from Lake George. The rocks here are chloretic slates, quartzites, conglomerates, &c. They belong to the Huronian series and contain veins of yellow copper ore.

Although extensive preparations have been made for mining and shipping the ore, I do not think up to the present time any profit has been derived from these mines.

A short distance from the mines at Limestone Point, a stratified dolometic limestone occurs in large quantities as on the banks of Garden River. This limestone is well adapted for ornamental purposes. The scenery in this vicinity is wild and rugged; it is a delightful spot for the sportsman. A pike weighing fifteen pounds was taken from the lake by my Indian guide. Returning to Sault Ste. Marie, I engaged two half-breed Indians and their boat and we started on a coasting journey down St. Mary's River, part of Lake George, along the north side of St. Joseph's Island, where at Gravel Point I secured a large number of fossils belonging to the Black River formation, and finally arrived at the Bruce Mines on the north side of Lake Huron, where at fifty fathoms from the surface, in 1847-9, 400 tons of copper ore was raised. But the working of these mines was discontinued many years ago.

At the time of my visit one of the pioneers—if not the discoverer of these mines—Col. Rankin was engaged loading a barque with the quartz debris from the stamp mills. He hoped to dispose of this broken rock to a Boston firm who were to use it for covering the walks of parks and private residences. There were then many thousands of tons of this material, but whether Col. Rankin made a paying speculation of it or not I do not know. Associated with these copper-bearing rocks are great bands of jasper conglomerates, which form quite a feature in the Huronian district. It is a beautiful rock, remarkable for its bright red jasper pebbles. Masses of detached pieces, some almost as large as a small log cabin were seen in or near the village of the Bruce mines. I shipped a large block of this conglomerate to the Survey, together with many other specimens characteristic of the Huronian formation, and on the 19th of August, returned to Sault Ste Marie where I again engaged two half breeds, and with a good boat, provisions, &c. we started on a ten days' journey along the North shore of Lake Huron.

It would be tedious to record all the incidents and adventures we met with in this long journey, most of which was made on foot with one man, while the other followed slowly with our boat,

AMONG THE ROCKS

landing whenever signals were made for him to do so. Some of our night camps were made on the small islands near the shore. They are composed of similar rocks to those of the coast, but in many instances are smoothed by the waters of the lake, affording a good opportunity to study the stratification of the strata. On one of these islands, composed of chloretic slate we found—although very much water worn, cavities made in taking the rock out with rude instruments, for (as my Indian guide said) making stone pipes. Each of my men carried away pieces for this purpose. They said this spot was known to their forefathers, who journeyed many miles to obtain this favorite " pipe stone."

Arriving at Thessalon Point we ascended the river of that name for twenty miles or so, passing over many of the typical rocks of the Huronian series. Returning we continued along the coast arriving at Algoma Mills Sept. 2nd, where my men were paid off. I saw them and their boat safely on board the steamer bound for Sault Ste. Marie, their home, and after spending a day or two on the large exposures of rocks in that vicinity, boarded the returning steamer for Owen Sound, and thence by rail returned to Ottawa which place I reached Sept. 6th, after a journey not free from dangers, and geologically, of less interest than any of the previous journeys I had made. A few days after my return from the above lakes I was again in New Brunswick working up palæontological evidence regarding certain formations then under discussion.

Returning to Point Levis a few days were spent among the Graptoletic rocks of the "Quebec group," and on the 5th of October, I once more settled down to my winter's work connected with the Survey, which consisted, as usual, in musuem work, mending and restoring a number of fossil bones collected by Dr. G. M. Dawson and Mr. R. G. McConnell. My microscopical work consisted in preparing 300 sections of rocks and fossils. A large number of specimens were developed and prepared for study and for the museum, and the remainder of my time was spent in arranging for my next summer's field work.

The duty assigned me for the summer of 1883, was to make a geological examination of that portion of the North-west lying between the Cypress Hills, Manitoba, and the Rocky Mountains in the vicinity of Kootenay, or Waterton Lake, a distance of about 500 miles.

As Dr. G. M. Dawson—now director of the Survey—accompanied by Mr. J. B. Tyrrell, also of the Survey was about to make a Survey of a portion of the North-west Territory this summer, it was thought advisable that I should accompany him as far as Fort McLeod, Alberta, near the line of the fifth principal meridian.

SHIPPING BUFFALO BONES, MAPLE CREEK, N.W.T., 1883.

AMONG THE ROCKS.

I left Ottawa, May 22nd, picked up my son G. H. Weston, then a student at the Agricultural College Guelph, Ont., who was to accompany me as assistant—and proceeded to Toronto, there to wait the arrival of Dr. Dawson, who joined us a few days later. We left Toronto, May 27th, and started by rail via Hamilton, Chicago, St. Paul &c., arriving at Winnipeg on the 30th, and Brandon the same evening, where we purchased part of our travelling outfit and chartered a freight car which for the remainder of our rail journey became a home not only for our horses, waggons, hay, &c., but for Dr. Dawson, Mr. Tyrrell, myself and our assistants.

The construction of the Canadian Pacific Railway was then in progress, and in the vicinity where we were, the one train running on this section consisted of construction cars and one construction passenger car. We got our car attached to this local train and started towards Swift Current. The first town we struck, after leaving Brandon was called Red Jacket. It consisted of two large canvas tents, one a general store, and a sign on the other stated that it was " Red Jacket Hotel."

We reached Maple Creek June 9th, where we found Mr. R. G. McConnell of the Survey, camped a short distance from the Mounted Police Station. This was the end of our railway journey, and here we pitched our tents and completed our arrangements for the journey across the plains.

Long before the great Canada Pacific Railway passed through that portion of the North-west Territory known as the Assiniboine district, Maple Creek, was one of the principal Indian posts. Here Crees, Assiniboines, Sarcees, Bloods, Blackfoots and other tribes brought their furs and other produce and traded with the white men. Although most of the land of this district is now owned by the white settlers it is still the favorite camping ground of the aborigines.

A mile or so from our camp were congregated a large number of Indians belonging to various tribes. In their tepees, we were told, they were preparing for their annual "Sun-dance."

Thinking this a favorable opportunity to get a few good photographs we caused one of our vehicles to be hitched up, then Dr. Dawson and I drove across the prairie towards the wigwams. When within half a mile of this camping ground we noticed that close to the tepees were probably a hundred Indian ponies already bridled. A moment later, as many bucks in all their war paint and feathers issued from their tepees, hastily mounted their horses, formed a line and started at full gallop in a direct line for us. On they came till within a hundred paces or so of us. For a moment I sat paralyzed, expecting to be trampled under their horses'

AMONG THE ROCKS.

feet, but when within thirty or forty feet of us they opened out in the centre allowing room for us to pass, and with a rush and the Indian salutation "How," they were soon far away. It was a beautiful sight, one not soon forgotten. Arriving at the Indian camp, Dr. Dawson attempted to get a photograph of some of the squaws and young Indians, but the moment we got the camera up, away they scooted into their tepees. However I seized the hand of an almost nude youngster, and he and I form part of a pretty picture (?)

Both inside and outside the general stores of Maple Creek at this time, 1883, were favorite lounging places for the Indians. Many of the aborigines were quite picturesque looking, adhering strictly to the Indian costumes of their fathers, while others adopted any of the white man's cast off clothing they could find. The respect accorded an Indian appeared to depend on how many ponies he possessed and his ability to support himself without the aid of the white man. In this district there appeared to be a great many Indians who absolutely refused to remain on the Government reserves, therefore obliging themselves to pick up a precarious living, one method being the collecting of buffalo bones for which so much per hundred pounds weight was given by men commissioned to receive and ship them to various towns to be used in refining sugar, fertilizing, etc. It no doubt seems hard to those who a few years ago hunted the buffalo in all its wild grandeur to be gathering and selling for the small sum of money or provisions the bones which in their happy hunting days they left to bleach on the plains.

By various signs I talked to an old Indian—whose photograph now adorns a wall in the ethnological room of the geological museum—of the glorious days of the buffalo hunt--could one have photographed the various expressions of this aged buck's face and the beautiful way he illustrated—by signs and gestures the buffalo chase, and later the total extinction of this noble monarch of the plains—it would have been a picture indeed.

We had completed our arrangements and proposed leaving the following morning, June 6th. Our horses eight in number, two heavy waggons, one buggy, and one buck-board were all arranged round our tents. It was late when we left our camp fires and tumbled in for the night. At 6 o'clock next morning Thompson—one of Dawson's men—reported that three of our horses were missing. We had been warned by the mounted police that Indians were stealing horses in all directions. This made us anxious and men were at once sent in various directions to look them up. It was dusk in the evening when my man Haultain returned with our horses,

AMONG THE ROCKS.

which he said he found among a number of Indian ponies miles away. Dr. Dawson had brought with him a branding iron, a broad arrow, the government brand mark, and each horse was at once branded. A sharp look out was kept during the night and the following morning after an early breakfast and a good-by to our friends of Maple Creek we started on our journey towards the Cypress Hills. Our mode of procedure was: Dr. Dawson and his assistant, Tyrrell, in a buggy, Johnson on horse back, who chose the best road for our waggons, then followed their supply waggons. Next came my equipment waggon, and my son with a buck-board, and lastly myself on horseback.

In five days we had reached and crossed a portion of the Cypress Hill and were camped in a valley about 150 miles from Fort McLeod. The discovery this year, 1883, by Mr. R. G. McConnell of the Geological Survey, and later by myself, of mammalian and other fossil remains, had rendered the Cypress Hills' district a most interesting localities for the geologist and osteologist. But I shall speak of the rocks and fossils of the hills and valleys of the Cypress Hills in my notes of next summer's travels.

The 12th of June found us in the vicinity of Pagon Creek on the Fort Walsh trail. Here we gathered as many mushrooms as we wanted. The scenery here is very fine. By compass we had travelled over hills and through valleys where we had found both wood and water, but now we are about to journey for a few days through a district where wood is more scarce than the "Prairie dog," *Spermopoilus Ludovicianus*, and our men are filling every available space in the waggons with food for our camp stoves.

As we travel these plains I can't help thinking of the time—not many years ago—when vast herds of buffalo followed each other, in single file, along the narrow trails we so often cross, which if we follow, we will surely come to water. Then the Indian was perfectly independent of the white man, for the buffalo furnished him his clothing, his tents to live in, and meat to eat. But all is changed now, bark and dirty rags form their wigwams, and those who will not remain on the reserves, furnished them by government, must eke out a living as best they can. The small cash treaty annuities (which I think average from $3 to $25 according to age and rank) is frequently spent the day it is received, generally on a few luxuries for the buck, not for his wife and children.

While passing through the beautiful valleys of the Cypress Hills, and over the Sweet Grass Hills, we followed the advice of the Mounted Police and kept a sharp look out for Indian or other horse thieves, by having our men take turns in mounting guard during the night. The horses

AMONG THE ROCKS.

were picketed near our camps and each man, well armed, took a two hours' watch.

Many mishaps incidental to travelling the prairies were met with—a break down while crossing a soft-bottom stream, an axle-tree broken, through a wheel dumping in a badger hole and so on. But the beautiful scenery and the important geological facts gleaned as we went along amply repaid us for the cares of the day, and at night when safely housed in our tents with the spoils of our journey consisting sometimes of Indian skulls, fossils, &c., new food for the osteologist and palæontologist, we felt that we had done some little for our "Queen and country". The chief incidents of interest during the next few days were the discoveries of small seams of coal (lignite); rocks with beds of fossil oyster shells, small herds of antelope; water which when separated from thousands of small red insects by straining through a handkerchief, was not so bad, no wood and a scarcity of buffalo chips (dung) for our camp stoves, a constant look out for redskin horse thieves, and many other things which tended to impress this section of the Territories on one's memory.

Passing through Verdigris Couleé and along the margin of Suds Lake we arrived at Milk River Ridge in the vicinity of which a remarkable fossil fauna was found. It is probable that fifty different species were obtained from a patch of rocks a couple of yards square. We named this spot Fossil Couleé. It can be found in Dawson's map of this district. This unique collection together with all specimens obtained during our long journey from Maple Creek, now lies at the bottom of Lake Superior owing to the sinking of the ship Glenfinlas. It is fourteen years since this occurred, but I still mourn the loss of these hard-earned specimens which were shipped by bull-team from Fort McLeod to Maple Creek.

From the top of Milk Ridge we got our first view of the Rocky Mountains. On the 25th, of June we struck the Benton trail and saw one of the interesting sights of prairie travel, a train of twelve great schooner waggons divided into four sections, each section having six teams of mules It is worth a long ride to see one of these caravans cross a soft-bottom stream, especially if any of the harness breaks and the mules get mixed up—then one hears language which—well I certainly should not like to have a member of the fair sex with me at that time. The mule driver is swift in overcoming difficulties, but though his poor dumb companions may strain every nerve to straighten things out again, they never go unpunished for any mishap that may occur, and at such a time it is well for the tenderfoot not to interfere.

Passing Ed. Mahan's couleé we camped for the night at fifteen mile lake, within a short distance of a small encampment of Blood Indians who were not long in paying us a friendly visit, shaking hands all round and saying "How"— We presented each with a slice of pork and while Dr. Dawson was fixing his camera to get one of his historical pictures I presented each one with half a plug of black tobacco, but at the words "all right, steady," from Dawson, off they went making a bee-line for their wigwams.

June 27th, we reached Belly Valley through which the swift and often turbid Belly River courses. To our right are drift deposits forming high steep bluffs under which are extensive beds of coal which were being worked. The opposite side of the river is Coal Banks, so named from the coal deposits. The river here is said to be 437 feet wide; the prairie level 300 feet above the stream, and the scarped banks—which present beautiful sections of these stratified rocks—200 feet high. The rocks of the Coal Bank district belong to that portion of the Cretaceous formation known as Pierre shales, Belly River series, &c., some portions of which are almost destitute of fossils, while other beds are exceedingly prolific in organic remains.

After days of travel over woodless plains, it is exceedingly pleasant to strike a verdant spot like Belly Valley where wood is plentiful and the blossoms of the thousands of wild roses perfume the air—but, tread carefully, for the deadly rattlesnake lurks among the plants.

One would have liked to linger many days in this charming spot, but having secured some baking powder, a few other luxuries, and two or three good photographs, we crossed the stream on the Coal Banks ferry for which service we paid $4.75.

The following morning we left Coal Banks, climbed the steep hill and were again on the prairie level; continuing our journey we passed Rye Grass flat, of which I shall speak later on, and camped for the night on the banks of the Old Man River where a large number of Indians had pitched their tepees, and appeared to be holding a pow-wow.

The following morning, after a journey of eight miles over an excellent trail, we arrived at Fort McLeod, one of the important trading posts of the Alberta district. Here all our precious rocks and fossils—which I have already stated now lie at the bottom of Lake Superior—were repacked and left with J. G. Baker & Co. for shipment by the first bull or mule train which left for Maple Creek.

Lieut. Gov. Dewdney arrived at this time and was given a good reception by the few white men of the village and a dozen or so of Indian bucks who came from—well I don't know where, but I do know that among them seated astride on an Indian pony was a fair (?) daughter of the plains who cast sheep's eyes at our worthy Lieut. Governor. "Nothing venture, nothing have" was quoted by one of our men.

How shall I describe this place where dirt prevaileth and righteousness entereth not within her gates—well perhaps it is better to leave that to a more prolific pen than mine.

Sunday, July 1st, and the two following days we had a busy time getting horses shod, waggons mended, laying in provisions, &c., preparatory to our journey to Kootenay Lake, about 150 miles west.

So far Dr. Dawson and his assistant Mr. J. B. Tyrrell, had been our guides and companions, but on July 3rd, they left us and started across the plains in an opposite direction to the one we had to take.

Any one who has travelled with Dr. G. M. Dawson will I feel sure look back with pleasure to their journey with that most courteous gentleman whose faculty for overcoming difficulties, and great knowledge, has won him the high position of Director of the Geological Survey of Canada.

We watched the doctor's outfit till it was lost in the distance and a few hours afterwards the "Weston outfit" was heading for the Rocky Mountains. It consisted of one heavy waggon drawn by two fine strong chestnut horses—

REMINISCENCES

"Baby, and Dick;" a buck-board with a yellow horse, "Buck," and my saddle-horse—"Pink eye." Haultain our driver and my son G. H. Weston, camp equipment &c., for a journey of six or eight hundred miles over the plains. As on these journeys one's life often depends on one's horses, it is not surprising that they are well looked after and that their various peculiarities are noted. Continuing our journey, sometimes by trail and sometimes by compass, we reached Pincher Creek, an important fossil locality, July 9th.

After leaving Fort McLeod, we journeyed over hills and through valleys, forded rivers and creeks. The high cut banks of the rivers in this locality often render it necessary to travel long distances before a crossing can be found, and then to cross these swift streams required a steady head. On one occasion Haultain had got his waggon safely over a broad swift river, and my assistant George on his buck-board was anxiously watching Buck's legs to see if they moved, but turning for a second to see how I was getting along, and seeing that my horse was turning round, shouted, "for God's sake let go your reins;" I did so, got a tight grip on the saddle, closed my eyes, and in a few moments felt my horse scramble out on the bank of the river; the rushing water had made me giddy and I was keeping a tight rein on my horse, but the moment

AMONG THE ROCKS.

Pink Eye felt his head free he made for his companions on the shore. Since this incident I have often trusted to my horse, feeling sure his brute instinct would overcome what to me were grave difficulties. Halting for our noon meal about fifty miles from Fort McLeod we discovered that one of our dunnage bags was missing. Haultain was just about to return on our trail to look for it when in the distance we saw four wild looking Indians mounted on bare-backed ponies. One horse carried our dunnage bag tightly lashed on his back. The day was very hot, and for the journey the Indians had divested themselves of almost all their clothing displaying a good portion of their forms. By a diagram drawn in the sand they reported that the bag had been found in one corner of the Indian reserve, and by signs demanded $2 each. I showed them the government brand on our horses; we made them understand that the bag was the property of the government and that fifty cents each was the price of the work they had done; after a little parley with each other they decided to accept our terms, shook hands, said "How," mounted their horses and at full gallop started back for the Blood Indian Reserve.

Reaching Pincher Creek, July 9th, we pitched our tents in a picturesque locality on the bank of the creek, a mile west of the North West Mounted Police farm. Near us were camped two families

REMINISCENCES

of West Percy Indians. The bucks were fine strong dark skinned fellows who during our two or three days' stay in this locality became very friendly, especially a stout well formed youngster who made his initial visit to our camp in a perfect state of nudity.

At Col. McLeod's house, a mile or so from our camp, we found our mail from Ottawa. It had been twenty-two days on the road, but was very acceptable.

The geology of Pincher Creek is very interesting, winding its way, as it does, through the Foot Hills of the Rocky Mountains, the cut banks of the creek afford an excellent opportunity to study the various beds of rock which constitute the Laramie or Eocene of Canada—either the base of the Tertiary or the upper beds of the Cretaceous. Some of the rocks here are very prolific in fossils. One cliff perhaps 200 feet high holds a thick bed of limestone partly made up of well preserved gasteropods and other shells. The beautiful spiral shell *Physa Copei* described by Mr. Whiteaves in "Contributions to Canadian Palæontology," Vol. 1, Part 1, is very abundant. Some of the shaly beds hold fossil plants. Red, purple and grey clays overlie the fossiliferous rocks.

We reached Mill Creek—another important fossil locality—July 13th, while a big hail storm prevailed. As at Pincher Creek, we succeeded in getting several good geological photographs of these localities.

The rocks, which are well displayed here, form a portion of the Upper Cretaceous formation. They have been divided into Mill Creek and Niobrara beds. The latter beds hold the well known bivalve *Inoceramus problematicus*, and the former beds, many beautiful shells, some of which have been described by Mr. Whiteaves.

Among the fossil flora of this locality are several species of ferns, and leaves of exogen plants. Some of these have been described and figured by Sir J. W. Dawson in Transactions of the Royal Society of Canada.

The Mill Creek sandstones of this vicinity are represented by high cliffs. In these we saw casts of large trees—relics of that epoch of this world's history when great carnivorous reptiles peopled a large portion of America— a time when crocodiles filled the waters and swarmed over the banks of rivers which now flow past the city of Philadelphia.

Our three days' sojourn in Mill Creek valley was a most pleasant one—surrounded by green pastures on which the finest cattle of the North-west roam. We had good trout fishing at the South Fork of the Old Man River; a grand view of the snow-capped Rocky Mountains; civilization in the way of a blacksmith's forge, carpenter shop, Gladstone's House, a boarding place,

REMINISCENCES

mill, white men, half breeds, Indians, and abundant food for the mind of the geologist and palæontologist.

It was not without regret that on the 16th of July, we pulled up stakes and continued our journey towards Kootenay Lake.

Following the trail Canada's Governor General, Lord Lorne, travelled over in 1882, and the Kootenay River, after many difficulties—fording streams and creeks, in which a broken axle of our waggon figured conspicuously, as it obliged two of us to return to Mill Creek; thence on horseback over rivers and hills to Garnet's farm where Dr. Dawson had left a makeshift axle which his men made from a stick of hard wood then carried swung under his waggon. This my son George brought down the hills, and over rivers and creeks, on horseback—no small task; which it gives me pleasure to record.

After a good night's rest at the creek, we returned with our precious stick of maple and some telegraph wire, to obtain which George and I had driven eighty miles, and with a broken shafted buckboard had twice forded the swift waters of Kootenay River. But at last we were again at the scene of our wrecked waggon, where Haultain had been guarding our provisions from prairie wolves and prowling Indians.

At daylight the following morning the wheels of our waggon were firmly lashed on our ama-

AMONG THE ROCKS.

tear axle, a new shaft for our buck-board was roughly made from a stick of cotton-wood cut in a small bush miles away, and other repairs were attended to. Our present difficulties being over, we resumed our journey and in two days camped in the vicinity of Kootenay Brown's shanty at the south end of Waterton (Kootenay) Lake, a magnificent sheet of water nestling at the foot of grand, rugged, picturesque mountains, some of which have an altitude of 4000 feet above the lake.

I did not see any fish from the lake, but from the home-made fishing tackle seen in Brown's house I should fancy some of the largest fish found in our Canadian Lakes inhabit these waters. Brown appeared to be an educated American, living with an Indian woman. Why he chose to become a squatter in this lonely spot, he refused to reveal to us. The heads and skins of the grizzly bear, mountain sheep and other wild animals which adorned his humble dwelling proved that he was no mean hunter.

With him as our guide I visited the "Cascade," a beautiful waterfall on the western shore of the lake. Many of the characteristic rocks of the mountains are well displayed here, and I hoped to obtain fossils from some of the beds, but my researches were fruitless.

The following day Brown was engaged to accompany my son and me through a portion of the South Kootenay Pass—a lovely valley through which a small stream of water courses. It was here while my horse was picking his way over a rough trail at the foot of high cliffs that Brown shouted "look out for the grizzly". Looking across the valley we saw a full grown grizzly slowly piloting her cub down the side of a pine clad mountain. Being unwilling to take any risks with her ladyship I moved off as quickly as possible, but Brown who was fully prepared for any emergency begged me to allow him to follow up his big game. When George and I were half a mile or so away, we heard the report of his rifle, but later on when he overtook us he reported that just as he fired his horse baulked, and as the bear charged at him, and his horse not being accustomed to that work, he thought it best to retreat. He said however, that he should renew the chase next day as he had an order for a grizzly cub, and if he could despatch the mother he could readily bag the young one.

The sun had set, when after a horse-back ride of forty miles or so we returned to camp. We had followed the narrow trail which for hundreds of years had been traversed by the Kootenay and other tribes of Indians when going to their buffalo hunts.

The magnificent ranges of snow capped mountains—the home of the mountain sheep and goat — which tower thousands of feet above the thickly forested sides of portions of the valley, the home of the grizzly bear, render the Kootenay Pass a most interesting spot for both the artist and the sportsman.

In wishing us good-night, Brown said " Keep up a good fire, boys, for there are more than two grizzlies round these diggings."

The geological horizon of the Waterton (Kootenay) Lake rocks, owing to the absence of fossils, has not been determined definitely, but in Dr. Dawson's survey of this district, he divided them into several zones—Cambrian, Carboniferous, Devonian, and probably Triassic.

Among the great variety of rocks which form the magnificent elevations of the Waterton Lake and South Kootenay Pass district—one of which reaches a height of 10,535 feet above the sea—are magnesian limestones, slates and shales, quartzites, amygdaloidal traps, etc., rocks of every tint; but fawn, bright red, yellowish and brown colors predominate. Sheep Mountain, one of the Kootenay Lake ranges which was measured by Mr. R. G. McConnell, of the Survey, who has done so much valuable exploration work in the Northwest Territories, has an elevation of 1,955 feet above the lake. It is composed of sandstones of various tints, red and green shales, jasper conglomerates, limestones, etc., some of the sandstones are ripple-marked. The south Kootenay rocks represent a thickness of about 4,100 feet, but in all this series of strata no fossils have yet been discovered.

Our geological work being finished, as far as time would allow, in this district, we struck camp July 25, and the following day forded the Kootenay River. Heavy rain and the melting of snow on the mountains had swollen the river so much that crossing became difficult—my horse lost his footing and swam for it, landing half a mile down the river. George thought Buck towed his light vehicle over, Haultain's waggon being heavy, did not float.

By July 27th, we were again at Fort McLeod, where we got a new axle for our waggon and a shaft for our buck-board.

While camped near the village waiting for repairs to our outfit, etc., we saw a good bit of the Indian life of this district. The Blood Indian reserve being only a short distance away, Indians were constantly travelling between the reserve and the village. Although associating with the white man, a great number of the Indians of this district still adhere to the paint and feathers of their forefathers. Many of the young Indian women of this place appear to be as vain as some of their white sisters in their adornments. But this is a mounted police settlement!

From sunset till the small hours in the morning we heard the monotonous tum, tum, tum, on the Indian drum, which seemed to accompany all gambling, the nightly pastime of the red man of this village.

In the corner on the mud floor of one of the Indian shacks lay a mother and her babe sleeping, while close by squatted two "bucks" deep in an Indian game of chance, One of the men was stripped almost to his breech-cloth. He had staked his blanket and lost; staked the few miserable household possessions, and lost; and still the play went on, and the half clad Indian boy sat in the corner and beat his tum, tum.

. . . .

July 30th, we left Fort McLeod, and camped for the night at Willow Creek, and the following day reached Rye Grass Flat, Old Man River, a locality of much geological interest. Here thick beds of yellowish white limestone are crowded with well preserved fossil bivalve shells, which belong to the Lamellibranchiata family. Some of these forms have been described and figured by Mr. Whiteaves; Contributions to Canadian Palæontology, Vol. 1, Part 1. These fossils which are well represented in the cases of the Geological Museum belong to the Laramie, which forms the base of our Canadian Tertiary formation.

Leaving Rye Grass Flat—so called from grass of that name which is characteristic of this valley, Aug. 3rd, we proceeded by compass across the prairie level to Scabby Butte, where we expected to camp for several days, but on our arrival we found the small lake of that locality dried up, and after vain attempts to get water, even enough to make a cup of tea, we decided to hasten back to our old camping ground, the only place for miles where we could get water.

A few days later we were again at Coal Banks, following the banks of the Belly River. High scarped yellowish sandstones form an interesting feature of this locality. Besides holding many important fossil shells, we found several bones of a great Dinosaurian reptile, probably the *Lælaps incrassatus* (Cope). The prairie level here is probably 200 feet above the river. In one or more instances we found much difficulty in getting our horses down to the water, and several times had to carry water for camp use half a mile or more up these steep banks.

Aug. 12th, we were following the banks on the north side of the St. Mary River, which enters the Belly River a few miles above Coal Banks. It has its source in the mountains south of the 49th parallel, is exceedingly rapid, and passes through a beautiful verdant valley, in places 200 or 300 feet below the prairie level.

The rocks which belong to the Upper Cretaceous and form a portion of the Pierre and Fox Hill formations, contain many interesting fossils, but as yet, considering the extent of these formations, very little has been done regarding the palæontology of these great exposures.

AMONG THE ROCKS.

I must pass over our wanderings for the next few days and take the reader of these notes once more to our old camping ground at Rye Grass Flat, in the valley of Old Man River. We were returning to Fort McLeod and remained here to pay another visit to Scabby Butte. It was the 15th of August. The water in all the rivers was low and the ponds and swamps dried up. Fearing we would again find the Scabby Butte district destitute of water, I decided, with George, to make the journey with one horse and a buckboard. So with a large tin pail of water, for our horse Buck, and a small keg of water for ourselves slung under the buck-board, we started at 7.30 a m., and following our old track reached the Buttes in about three hours. Scabby Butte is a remarkable place, reminding one of the excavations of an ancient city. The sands, clays, lignite shales, lignite coal, sandstone and other rocks, have been scooped out, as it were, from the prairie land, to a depth of fifty or sixty feet, leaving portions of stratified rocks standing up like the walls of a burned city. There is evidence that at one time this was a great rendezvous and hiding place for Indians during their buffalo hunts and probably during their time of war. The geological horizon is Laramie, supposed Lower Tertiary, or Eocene. Some of the same fossil shells found at Rye Grass Flat, and other localities mentioned, occur here, but the

REMINISCENCES

chief interest of Scabby Butte lies in its fossil bones. It was the home of the great Dinosaurian reptiles.

> . . . Monsters of the prime,
> Who tare each other in their slime.

Mr. R. G. McConnell, of whom I have several times spoken, was the first to discover, in 1881, fossil bones at this locality. In the year mentioned he dug from a soft sandstone cliff, a femur or thigh bone of one of these extinct kangaroo-like reptiles. This bone is about the size of the thigh bone of a mastodon. It fell to my lot to restore it, for it was in many fragments.

Another interesting feature of Scabby Butte is the fossil wood found in the sandstones and other rocks. Many feet below the prairie level we saw the stumps of exogenous trees—trees which grew in the silent ages of this world—silent but for the struggles of fierce reptiles, whose sharp cutting teeth show that they were flesh eating animals; and for the sound of restless waters which have left us as souvenirs examples of ripplemarkings. Here, too, these large trees spread their branches over delicate ferns and grasses which have left their impressions in the sandstones and shales. The fossil wood of this, and many other places in the North-west, is of much palæontological value when it is found in place, as it grew. Being silicified these

AMONG THE ROCKS.

woods—stone woods they are frequently called—retain their cellular structure. This enables the palæontologist to determine their species.

Sir J. W. Dawson has described in Trans. of Royal Soc. of Canada, from sections made by the writer, a number of these woods. They show that the flora of those days contained, poplars, pines, oaks, and other trees which grow at the present time.

One of the specimens of fossil wood we brought from Scabby Butte is about five feet long and a foot wide. It shows it has been very much flattened by the pressure of overlying beds, rocks of great thickness which must have overlain the present beds and been denuded before the time of the present prairie level.

The sun was getting low when Buck—who had refused to drink the water we had taken so much trouble to carry for him—was released from his picket rope, and in a short time we were retracing our trail over this trackless portion of the prairies. The only animal life we saw while on our Scabby Butte journey was a small herd (six or eight) of antelope and some blackbirds, one of which for a more elevated position perched on Buck's head, and on this perch Mr. blackbird got a free ride for several miles flying off at times, then returning. But this is not an unusual sight on the prairies. We reached camp in the gloaming, where we found

REMINISCENCES

Haultain prepared to give us a good supper of prairie chicken and pork.

Continuing our journey from Rye Grass Flat, past Coal Banks and through a part of Chin Coulee—over a portion of the country where long distances have to be made between drinkable water—we reached the South Saskatchewan River, a little below the confluence of the Bow and Belly Rivers, August 23rd. The valley here, which is 250 feet deep, exposes a fine section of Cretaceous rocks. Many of the beds are prolific in fossils especially the genus *Ostra*. Some of these oyster beds are very interesting as they contain fragments of bones which probably belong to the Dinosaurian reptile, *Lælaps*.

The current of the river here is swift, and the water, which is beautifully clear, of an aquamarine color.

We pitched our camp on a flat a little below the Forks of the Bow and Belly rivers. The vegetation was chiefly wild sage, grass, and sunflower plants, which latter at this time where in full bloom, looking from a distance, like a cloth of gold. These plants formed an excellent covert for prairie chickens which were quite numerous.

Before leaving this interesting but lonely spot, George and I visited an Indian cairn—a few miles below the Forks of the Bow and Belly rivers—and while returning to camp in the evening saw, on the opposite side of the banks

AMONG THE ROCKS.

of the river, a man on horseback, making signs which indicated that he was lost and hungry. We signalled him to follow the banks of the river towards our camp, but he evidently did not mean to lose sight of us for he dismounted and at once commenced to lead his horse down the 250 feet of shelving rocks—a feat one might suppose only a goat could accomplish. We watched them with almost abated breath expecting every moment to see both horse and man roll to the foot of the cliff. But no, they landed safe at the water's edge where on a small grassy flat the man picketed his horse and then followed the stream—keeping a sharp eye on us as we followed the banks two or three hundred feet above.

In due time we reached camp, but how to get our man over a river too deep to wade and too swift to swim we did not know, till Haultain said he had seen up the river a roughly made weather-beaten boat—left there probably by surveyors, years ago. This we decided to try and use, and as there was no stick of wood we could employ as a paddle down came our tent, and with the poles for paddles, and partly stripped for any emergency, George and Haultain started across the stream, but before they were half over the current almost swamped their leaky boat. However they landed safely, and with the lost and hungry man, and a short slab of wood they had found where they landed, to assist, they crossed the river again, landing a mile below our camp.

We found that our visitor was a butcher from Medicine Hat, who had been hunting stray cattle, and had been without food for three days. Our supper, which was already waiting us, consisted of boiled prairie chickens, pork, beans and hard-tack biscuit, of which Mr. Butcher ate till we were obliged to stop the supply as we feared he might "overdo the thing" as Haultain said.

In a short time the stranger was fast asleep and next morning after a good breakfast, and with three days' provisions and a sketch of the river he was to follow, he crossed the stream again. The winding of the shore prevented us from seeing where he landed or how his horse and he succeeded in reaching the prairie level again.

This man—whose name was George Spindle—stated that within five miles of our camp Indians were hunting buffalo and two days ago he had seen more than one, and would have chased them but for the want of cartridges for his rifle—which we supplied him with before he left us. I am doubtful however about his having seen buffalo, as we had seen no fresh tracks of them on all our journey.

Leaving the Forks of the Bow and Belly rivers we continued our journey along near the top of the scarped banks of the South Saskatch-

AMONG THE ROCKS.

ewan which in places here is 100 feet wide and runs through a valley 250 or 300 feet deep.

A prairie fire had lately swept over this portion of the country leaving 15 or 20 miles of land destitute of grass. We descended the banks of the river when an opportunity afforded, and while our horses fed, made our geological examinations; always finding interesting palæozoic objects—now the scale of a fish, which must have been of great size ; now fragments of bone belonging probably to our great Dinosaurian reptile, now fossil shells belonging to the oyster family, and in the shaly beds, fragments of grasses and leaves—relics of the Cretaceous epoch of our world's history.

We reached Medicine Hat, Aug. 28th. "The Hat" is one of the live prairie towns of the Northwest—an important station on the Canadian Pacific Railway which crosses the South Saskatchewan River a little below. At this time (1883) most of the shops and dwellings were of wood. Besides twenty or more merchants there were two doctors, two barristers and solicitors, four hotels, and several boarding houses.

A large flat in front of the "town" afforded an excellent camping ground for Indians whose tepees numbered a hundred or so. While some of the Nitches—as they are called here—pick up a precarious living, selling to the white man any fancy ornaments they may be able to make, a few of the "gentle sex," with painted faces and gaudy Indian attire, solicit favors from those men who have no regard for either body or soul. And a Salvation Army man stood at the corner of the street and beat his big drum.

Rattlesnakes must not be uncommon in this section of the country, for several of these deadly reptiles were exhibited in the shop windows of the town; one was in a case outside the C.P.R. ticket office. On a subsequent visit to this vicinity I had the pleasure (?) of almost treading on one of these creatures, and should have done so, had his snakeship not sounded an alarm with his rattles.

Many interesting memories are associated with our short stay at Medicine Hat. Here we packed and shipped several boxes of specimens for study and for the museum. One of the boxes contained a fine bleached skull and other buffalo bones from one of the many "buffalo graves" we had seen along our line of travel. We had now completed our geological work, and at 3 p.m. Aug. 29th, started on our last prairie journey. We made 17 miles that evening before camping for the night. Our horses seemed to know that they were nearing their home, for they were very restless all night. We were greatly annoyed too by the howling or barking of prairie wolves. With great care we kept our

AMONG THE ROCKS.

horses till early morning, when after our last breakfast on the plains, we left for and arrived at Maple Creek, which place we had left on the 8th of June.

During our almost three months of prairie camp life I had ridden horseback many hundreds of miles over the barren and fertile fields of our Canadian North-west, chased the antelope and coyote, seen that important institution of prairie ranches, the cow-boy, in his dangerous and exciting work, "rounding up," visited the Indians in their wigwam, etc., and what was more interesting to me, secured many valuable specimens which have found a resting place in the Geological Survey museum of the Dominion. After storing part of our camp outfit and other government property we left for Ottawa which place we reached Sept. 10th.

CHAPTER IX.

TO SOUTH JOGGINS, N. S.—PALÆONTOLOGY OF THE CARBONIFEROUS ROCKS—JOURNEY TO MANITOBA AND THE NORTH WEST—MAPLE CREEK AND THE CYPRESS HILLS—VERTEBRATE REMAINS—A HERD OF WILD CATTLE—BELLY RIVER AND THE PRAIRIES—SOME INDIANS—SELKIRK AND ITS ATTRACTIONS—A PELICAN—THE RED RIVER—WINNIPEG AND ITS MISCELLANEOUS STORES.

SIX days after my long North-West trip, I was on the road to the South Joggins, N. S., where I again worked along the coast gathering new information regarding the palæontology of the Carboniferous rocks. I returned to Ottawa, Oct. 5th, with many fine specimens. My exploration expenses for this year, 1883, were $1321.46.

The winter of 1883-84 was as usual fully occupied in office and museum work.

The preparation of "Contributions to Micro-Palæontology" by A. H. Foord, artist and assistant Palæontologist to the Survey, required a great number of micro-sections of fossil Polyozoa, or Bryozoa. This interesting work took up a good portion of my time, and the remainder was spent in preparing fossils for study and for the cases of the museum, writing labels, superintending the mounting of specimens, &c. &c.

To defray my expenses during the summer 1884, I received from the director of the Geological Survey of Canada a letter of credit for $1,700.00 with instructions to proceed to Manitoba and the North-West Territories to collect fossils and obtain all the geological information I could from the rocks of the Cypress Hills, and from the deposits on a portion of the south shore of Lake Winnipeg. *

As in other cases already mentioned I was to give another young member of the staff his initial exploration journey. This young student in geology was James A. Macoun, son of the veteran botanist of the Survey, Prof. Macoun, M.A., F.L.S., F.R.S.C.

The little I knew of James Macoun led me to believe he would make a most agreeable companion. " What about Macoun?" said the chief on our return to Ottawa ? "Give me James every time," was my reply " he knows no fatigue, and fears no danger."

With camp equipment which included a good Winchester rifle, a Remington gun, fishing tackle &c., we left Ottawa May 21st, and proceeded by rail to Toronto and Owen Sound, thence by S. S. Athabasca through a portion of Georgian Bay, Lake Huron, and Lake Superior to Port Arthur. The voyage over part of these two magnificent lakes, and along a large portion of Georgian Bay —which together have an estimated length of 635 miles, and a breadth of 350 miles—on one of

---

* Dr. G. M. Dawson the present Director of the Geological Survey of Canada has kindly allowed me to make use of my field note-books, for dates during the remainder of these Reminiscences.

the powerful and well equipped steamers of the Canadian Pacific Railway, from Owen Sound to Port Arthur, occupied from 8.30 p. m. May 22nd, to 12 p. m. May 25th. While in Georgian Bay we were detained some time by fog, and soon after passing through Sault Ste. Marie locks into Lake Superior, we passed through a large body of field ice. The nights were cold, and while on deck an overcoat felt comfortable. We had several opportunities of seeing some of the extensive fisheries of these great lakes, and of tasting white fish and salmon trout just after being taken from the water. The scenery en route is in many places—especially the range of various colored cliffs along the south shore and Thunder Cape—very grand. The gold, copper and other minerals taken from the rocks in this section of Canada represent many millions of dollars, and yet it is thought that vast mineral wealth remains to be developed.

A motley group of people met our steamer at the Port Arthur landing; rough miners, Indians, wrapped in their bright colored blankets, beaded leggings and moccasins, the cow-boy with his broad brimmed hat, leather trousers and brace of revolvers stuck in his cartridge belt; the tourist with his photographing camera, and the man shark, " seeking whom he may devour." But we had little time to study this picturesque group, for after a hasty meal at the hotel we

AMONG THE ROCKS.

were soon in one of the handsome Pullman cars bowling along towards Maple Creek, a distance of about a thousand miles from Port Arthur—from which place, it will be remembered, we started on our prairie journey in 1883.

Arriving at Maple Creek, May 29th, we found the " town," which consisted of three stores, two hotels and a few scattered dwellings—in a state of excitement over a white man who had been killed by an Indian. This trouble, with the comforting remark that Indian horse thieves were numerous over that part of the country we were going to, made us feel a little anxious. To make the journey at this time the people of the village thought dangerous, and the sergeant in charge at the mounted police fort kindly offered to give us an escort, but we decided to risk the matter, and at once set about getting our outfit in order. As usual the J. G. Baker Company, who have stores in all the principal North-west towns—came to our assistance, and in a short time we were provided with a waggon, buckboard, four horses, two half-breed Indians (cook and driver) and a month's provisions. All being ready we left Maple Creek at 12 noon, May 31st, and struck off for the Cypress Hills, and our interesting fossil bone locality, which I have already stated was discovered by Mr. R. G. McConnell of the Canadian Geological Survey. We made 18 miles that day and camped near a creek in close proximity to a

REMINISCENCES

large encampment of Blackfoot Indians who were engaged in collecting buffalo bones to be used for refining sugar and for fertilizing purposes. Continuing our journey the following morning, partly by trail and partly by compass, we reached our field of research—48 miles from Maple Creek, Sunday, June 2nd—a great verdant valley extending many miles in a north and south direction. Here we pitched our camp within a short distance of where the water flowed south into the White Mud River and north into Swift Current. From this main valley coulées branch off at right angles and extend from a short distance to several miles till they are lost in the prairies. In these coulées are found the best escarpments of Miocene tertiary rocks; which although several hundred feet in thickness are seldom seen in vertical sections of more than fifty feet. The sides of the coulées slope at various angles up towards the table land, and are partly covered with scrubby brush, grass and wild sage (*Artemisiacand*), while in the bottom a stream of clear water flows, and willow, pine and other trees form a shelter for the antelope and other smaller game.

The Miocene rocks of the Cypress Hills plateau consist largely of gravel in which pebbles of quartzite from half an inch to a foot in diameter predominate. In all the overlying strata fragments of fossil bones, either fish, reptile or mammal have been found, but it is in the agglomerate bands that most of the vertebrate remains, now in the Dominion Museum, were obtained. This consists of (in places) a four-foot bed of yellowish sandy limestone which when treated with hydrochloric acid leaves a residue of grains of quartz and fragments of a variety of other rocks, angular and partly rounded pieces of rock similar to the matrix. Occasionally pebbles of red, black and other colored jaspers, banded quartzites, chert and porphyry are found; all derived from the Laurentian or Huronian mountains east of these deposits, and transported during the glacial epochs. In these rocks are entombed the remains of an extinct Rhinoceros which Prof. Cope named *Menodus angustigenis*. It is the largest hoofed animal yet found in Canadian rocks. It was our good fortune to discover, besides many vertebræ, limb and other smaller bones, the largest part of a cranium of this ponderous beast. Some of the teeth are almost perfect—great sharp-edged cutting molars. The nasal bone is ornamented with two horn cores.

SECTION OF MIOCENE TERTIARY ROCKS NEAR THE HEAD WATERS OF SWIFT CURRENT.

a. Superficial and other deposits in which, in a bed of yellowish-white silt 20 feet above the agglomerate beds (*b* of section), the teeth of an extinct deer, *Leptomeryx mammifer* Cope, were found. 30 feet.

b. Agglomerate beds, containing rounded pebbles of quartzite, jasper and chert. These are the beds from which *Menodus angustigenis*, *M. Proutii*, *M. Americanus*,

REMINISCENCES

Hemipsalodon grandis, and other genera and species were found. 4 to 20 feet.

   c. Yellowish sand with a few pebbles of quartzite. 1'—2".

   d. Fine conglomerates in which a few fish remains were found. 0'—6".

   e. Sand and thin beds of conglomerate. 5 feet.

   f. Conglomerates in which a large bone was seen. 2 feet.

   g. Sand and fine gravel. 2 feet.

   h. Gravel, loose pebbles varying in size from half an inch to a foot in diameter. 15 feet.

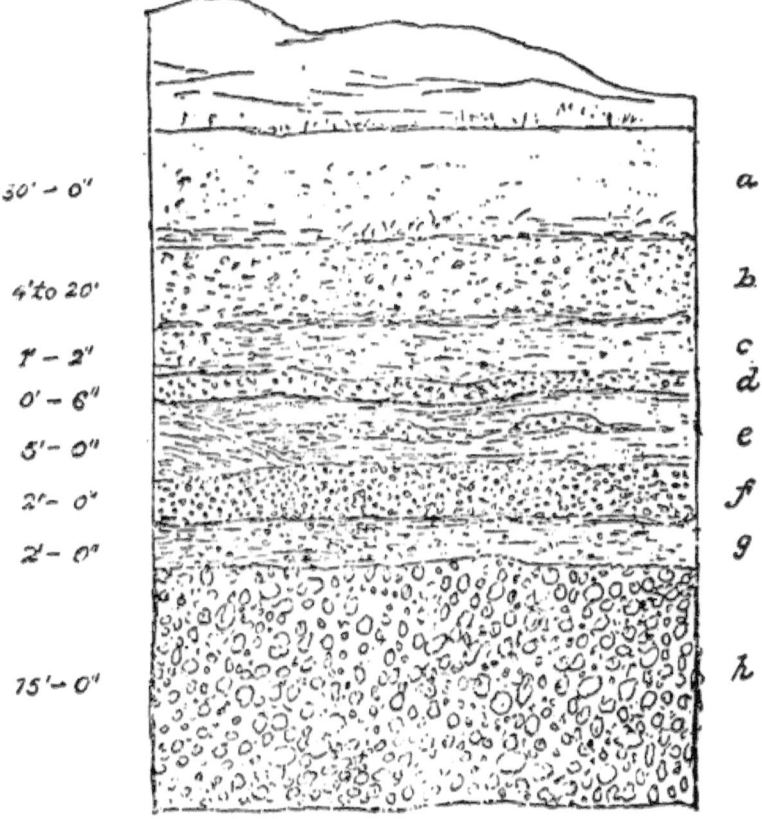

From Tran. of the Nova Scotia Inst. of Science, 1892-93.

AMONG THE ROCKS.

Another interesting specimen obtained is one of the rami of a large flesh-eating animal allied to the hyæna—a portion of one of the canine teeth of this powerful and ferocious animal measures one and a half inch in diameter. Then we found an almost perfect lower jaw of an extinct wild boar, *Elotherium arclatum*, Cope named this. Then we found a portion of the jaw of an extinct horse, which could not have been as large as a Newfoundland dog—*Anchitherium Westonii*. Prof. Cope named this after the discoverer and writer of these notes. As the remains of our little horse were found in the same beds with the remains of the great flesh-eating animal, it makes one think he must have had a tough time keeping clear of this savage beast.

Among other specimens from this interesting locality are the remains of extinct fishes, one bone of which Cope named *Amia Whiteavesian*, after the distinguished palæontologist of the Canadian Geological Survey. Then we saw and collected portions of large trunks of silicified trees, the lines of growth of which showed that they were exogenous and of great age. It is singular that in all these fossil bone beds, not a single specimen of a leaf or blade of grass was found. There must have been leaves to these trees, and if there were no grass or other vegetation, what did our little horse live on? Another

REMINISCENCES

singular thing is, that in this vast body of Miocene rocks only one cast of a shell was found, while the formations above and below are crowded with fossil shells.

By the numerous stone chippings and an occasional perfect arrow-head, we found that the pebbles of these Miocene rocks had been used by the Indians in forming their implements and weapons of war. A fine quartz arrow-head was found in one of the cut banks, two feet below the prairie level. How long did these two feet of earth take to accumulate over this relic of the red man?

In rambling through the great valleys and over the buttes of the Cypress Hills, one meets with incidents not soon forgotten. Out of many I could relate I must only mention one, for there is still much to record before we get to the end of our '84 travels.

Joe, my half-breed cook, and I were making a journey in our buck-board along the great main valley already mentioned, when on rounding a butte which had obstructed our view, we came face to face with a great herd of almost wild cattle. They were directly in our path, filling the space between the hills to our right, and the stream to our left. Had we been on foot we certainly would have been trampled to death. For a moment only, Joe hesitated, then grabbing a picket-line that lay at our feet, with tremendous shouts swung it around his head, while I at the same time urged the horse on It was a critical time, and to this day I can't think how we got through this herd of cattle who had yet to be subjugated by man.

Continuing south, we at last came to a place —in English—" The old man lying on his back." It was here during a war between two tribes of Indians a red man finding one of his opponents sleeping on his back, drove his tomahawk through his skull. Near this locality we came to what Joe was anxious to show me, viz:—A butte, composed—but I will quote from my notes on concretions, Transactions of the Nova Scotian Institute of Science, Vol. IX., Session 1894-95. " Another interesting concretion locality lies half a mile west of White Mud River, near the Fort Walsh trail in the Assiniboine district, N.W.T., the rocks belong to the Laramie formation. Here a small butte was pointed out to me by my half-breed Indian guide, who called the place gun-shot butte, and said, a few years ago when he, with others. hunted buffalo in that locality, they sometimes when ammunition was scarce used these balls in their guns and rifles. I found the hill or butte to consist largely of calcareous sand, which contained enormous quantities of spheroidal concretions, varying in size from that of buck-shot to an inch in diameter, the ordinary size being that of rifle balls, etc." These curious forms can be seen in the cases of the Dominion Museum. Near this spot we saw the remains of shacks which had been used during the buffalo hunt. "Ah!" said Joe, "they was the good times, good lot to eat and good lot of fun. But we must get back to camp before the sun he go down." So our horse's head, much to his satisfaction, was turned towards Skull-bone coulée, so named because of the fossil skull we had found.

It is known to all who have travelled the prairies with a small band of horses, how very "chummy" they become; so much so that if one is taken away, it requires a strong picket to keep the others from following; and how sensible they are to the approach of a companion! One day I sent Joe on horseback to bring me a specimen of rock from a far distant cliff. The sun was set and I was becoming anxious about their return, when after a time one of our waggon horses began whinnying. I looked in all directions, but could see neither man nor horse, and it was two minutes or more before Joe and his horse appeared, making their way over the top of a butte half a mile distant. How the horse recognized the approach of his companion half a mile away, and before he saw him, we shall never know.

During our journey back to camp we saw several Indian graves. Joe told me that some years ago many half-breeds and Indians had died in this valley of smallpox and were hastily buried. The graves were covered with stones to prevent the prairie wolf or coyote from getting at the bodies, but one grave had been dug up and the bones scattered around. I felt loath to leave these great valleys and coulées of the Cypress Hills where every day brought to light some relic of days when the savage hyæna, the bulky rhinoceros, our little horse and many other long extinct creatures lived and were never hunted by man. But we had spent all the time we could afford, and with our precious specimens and several excellent photographs of the rock exposures, etc., we started on our return journey to Maple Creek, which place we reached June 19th, and were soon busy repacking specimens for shipment, settling accounts, photographing typical Indians for the ethnological department of the Survey Museum, preparing for our next field of research, etc.

Maple Creek was a busy place just then—ranchers were in for supplies; cow-boys flirted with dark-skinned Indian girls, and the red-coated mounted police officers and men who had driven in from "The Fort" for the mail, added a warmth to the picture. But it was sad to see the poor Indians toiling in with their rude carts laden with buffalo bones. Thousands of tons of these bones have been gathered by Indians from the Cypress Hills district.

REMINISCENCES

And now, before starting on our journey down the Red River, I must say a few words about another very interesting fossil locality about fifty miles east of Maple Creek, and a few miles from Medicine Hat.

At 2.30 o'clock June 23rd, Mr. Macoun, Joe and myself boarded a freight train on the C.P.R. and started for Irvin Coulée.

At a lonely part of the prairie called Walsh, our train was delayed eight hours waiting for right of way. The only building there, was a log cabin put up for telegraph purposes. The operator kindly gave us some supper, and then we spread our blankets and tried to get to sleep, but the constant ticking of the telegraph machine and the knowledge that in a box a few feet from me was a live rattlesnake, prevented me from getting even the proverbial forty winks. This was a lonely place for a man to live, but the operator delighted in nature and the wild prairies and spent much of his time with his rifle and dog. The prairie wolf or coyote now in one of the cases of the Geological Museum was shot by him, and presented at my request to the Survey.

At 2.30 we boarded our train again and at 6 o'clock the same morning reached Irvin. The only house here was the railway section-house, a rough place with rough line-men, but we were glad to get breakfast here, then to spread our blankets on the floor and take a few hours sleep.

AMONG THE ROCKS.

Fortunately for us, a man with a yoke of oxen was here from "The Hat" and we got him to take our camp outfit two or three miles down Irvin coulée, our next place of research.

We pitched our camp on the banks of Ross Creek at the head of a verdant coulée near the line of the C. P. Ry. A great portion of this district is a sandy desert, with masses of soft sandstone interstratified with harder beds and in places thin beds of lignite shales. By the action of water these rocks have been worn into all sorts of grotesque forms. See photo. p. 46, C. Geological Survey Report for 1884. These rocks which belong to the Belly River series—already spoken of—contain fish and reptile remains. Mingled with the bones of extinct turtle, on the water washed sun-baked sands, we found many bony enamelled plates, or scales of the Ganoid, and teeth which may belong to the same fish. The most important geological specimen found in Irvin or Ross Coulée was the scapula of an extinct animal, probably a Dinosaurian. It was a bone three feet long, eight or ten inches wide and half an inch to one or two inches thick. I spent six or eight hours uncovering this bone, and Mr. Macoun walked sixteen miles—to Medicine Hat and back—to get glue or some other material with a view to preserving this specimen. It was like most of these fossil bones, cracked in all directions. Well, after all our trouble, while lifting it from its sandy bed, it fell into a thousand fragments, and now lies at the bottom of one of the great excavations in these soft sandstones. Still I mourn the loss of this bone.

The fact that the coulée in which we were camped was a favorite resort for prairie wolves, rattlesnakes, and other animals, gave rise, as we sat round our camp-fire to many wolf and snake stories. One told by Joe, our cook, was that a rattlesnake will never touch a pregnant woman. Both half-breeds and Indians believe this. We saw several wolves and one antelope in this coulée. In shooting at one of these, the end of my rifle burst, the break only affected an inch at the muzzle, and Joe was soon busy sawing this off with the bread-knife. It took three hours to do it, but it was accomplished as well as though we had proper tools for the job. This rifle, which hangs close to where I now write, reminds me of an incident I may relate. While camped at Maple Creek this summer a very fancy dressed Blood Indian came along, a thick shock of hair stood straight up from his forehead, the lower part was colored green and the upper part red; his dark brown face was striped with red and yellow paint; his bright colored blanket half covered his almost nude body. Joe, at my request, interviewed this dude, and found that in a few hours he was to be married. I at once set up my camera to get a picture for our ethnological room

AMONG THE ROCKS.

in the Survey Museum, but the moment Mr. Indian saw the camera away he went. Joe fetched him back, showed him a plug of tobacco and gave him the rifle to shoot me if the " picture maker " hurt him in any way. By these means I succeeded in getting a good photo. of this bridegroom elect, and also of Joe (who was almost as dark-skinned as the Indian).

While Macoun and Joe had gone in search of a vehicle—no easy task in this locality—I wandered alone among the sandy buttes and clays of the " bad lands " of this district. I had found, and was bagging some small fossil bones, when a slight noise caused me to look up, and there, fifty paces away, was a large hungry looking wolf. It did not appear to be the night-prowling coyote, but a larger and darker colored species—probably the dusky wolf. For a moment I did not know what to do, but remembering what Joe had said, I gave the biggest war-whoop this valley had perhaps ever heard, and to my delight Mr. Wolf turned and walked off. Towards evening my assistants returned with an ox-cart and driver. We camped that night near the railway track close to Irvin section-house. Here I paid off Joe and started him on his road back to Maple Creek. He had been a good man and I tried hard to persuade him to accompany us on the remainder of our journey, but he would not go so far from home.

REMINISCENCES

The following morning Mr. Macoun and I visited a very interesting locality, first noted by Mr. McConnell of the Survey in 1883. In my notes on Concretions found in Canadian Rocks—Trans of the Nova Scotian Inst. of Science, Vol. IX., 1894-5—I remark as follows:

"The largest concretions seen by the writer were found in the Fox Hill and Pierre shales and clays of the North-

CONCRETIONS AT IRVIN COULEE, N.W.T.

West; a very interesting exhibition of these giant forms may be seen three miles north of Irvin station-house on the Canadian Pacific Railway. Here huge boulder-like spheroidal and ovoid concretions once held in the rocks, but removed by denudation (probably in the Glacial epoch, for Glacial striæ are seen on some of the flat beds), stand out in bold relief, resting on the flat and up-turned edges of shale and sandstone; and on the top of one of them, about twenty feet high, an eagle had built its nest of buffalo bones and the roots of the wild sage, for want of a more elevated situation, which does not occur in this locality."

We found many interesting fossils here, among which was *Lingula nitida*; *Protocardia subquadrata*; *Liopistha*, and other species new to these rocks. Crystals of selenite (gypsum) are scattered over the sands and clays of these bad lands—lands which till only a year ago, when our great Pacific Railway wound its way across almost trackless prairies—were seldom traversed, and then only by the red man and wild beasts, and now in hundreds of places, where previous to 1883, the Indian pitched his wigwam and the prairie wolf roamed, one sees fields of grain and probably runs against a barbed wire fence.

We left Irvin July 1st, and the following day drove from Medicine Hat to the Saskatchewan coal mines (distant about eight miles). Mr. Lawson the manager received us kindly, and we remained at his house two days. This is a remarkable place, with great coulées and buttes, which contain thick beds of lignite coal.

From some of the sandstone and iron-clay stone we obtained many good fossils of both plants and shells. One of the most interesting fossil plants found here—at a spot where Mr. Lawson the day before killed a rattlesnake—has been described and figured by Sir J. W. Dawson in Trans. Royal Soc. Canada 1885. Sir William calls this pretty fossil leaf *Brasenia antiqua*, but by mistake gives Mr. Lawson credit for its discovery. The geological formation here, is Upper Cretaceous, part of the Belly River series. The photographs we obtained of these Coal mines, give a good idea of the country and also of the Lignite beds. We returned to Medicine Hat, July 4th, and spent several days examining the rocks of that vicinity. We reached Winnipeg July 14th, and the same day we landed at East Selkirk on the banks of the Red River 30 miles from Winnipeg. Here we found lodging at a small but comfortable hotel. The Village of Selkirk is about three miles distant from the bank of the Red River. The chief attractions of the place then (1884) were the quarries, where great loose pieces of limestone are dug out of the clay and shale and sent to Winnipeg to be used in the construction of some of the fine buildings of that city; the arrival and departure of C. P. Ry. trains, and the lime kilns.

The limestones of this locality and a good portion of the Red River Valley belong to the (Galena) Trenton and Black River formation. They are whitish yellow in color and contain a large fauna of fossils, many of which have been figured and described by Mr. Whiteaves of the Canadian Survey. The fossils obtained by the writer and Mr. Macoun, and later by Mr. Lawrence M. Lambe—artist to the Geological Survey—whose beautiful drawings adorn the

AMONG THE ROCKS.

Reports—form one of the important fossil exhibits of the Dominion Museum.

It was several days before we could obtain a boat and men, but at last we secured a good stout sail-boat, one elderly Cree Indian—who had spent most of his life at the Hudson Bay Post—one half-breed, who owned the boat, and one month's provisions. At 9.30 a.m. July 19th, we left Selkirk Village and drifted down the Red River—a muddy stream with a strong current, which in a few hours after passing through the large and handsome city of Winnipeg, empties its sediment-laden waters into Lake Winnipeg.

For some miles after leaving East Selkirk, the banks of the river on each side are dotted with small thatched cabins or cottages, and a few larger and more pretentious houses. Most of these dwellings belong to half-breed Indians, some of whom have good farms and fine cattle. After this, for miles before reaching the lake, the land on either side of the river is low and marshy, with a thick growth of rank grass, and rushes growing to a height of five, ten, and even eighteen feet. This swampy land appears to be the home of wild birds, especially the bittern and other members of the heron family. Camping ground is very scarce, and when found is— well like most Indian camping grounds.

We pitched our first camp on the shore of Lake Winnipeg, July 21st. It would occupy

REMINISCENCES

too much space in these notes and perhaps weary the reader to follow us from island to island, and over long stretches of water and the shores of this beautiful lake, so I will only relate a few of the many incidents still fresh in my memory.

Our second camp was made on Elk Island, near the north shore and thirty or forty miles from the mouth of the Red River. It is a pretty island, three miles or so in length, well wooded with birch, poplar, pine, &c. My men said the lake was well stocked with fish, among which are pike, cat-fish, sturgeon, whitefish, etc. I can vouch for the cat-fish, for Macoun, by wading in up to his waist and slinging a line and hook baited with pork, brought out some very fine specimens. I was surprised to find how different fish, just taken from the lake, tasted from those of our canals and lakes near cities.

Boulders of every description of Laurentian rocks line part of the west shore. The bedded rocks and the fossils they contain resemble those of the Hudson River formation of Anticosti.

After spending a short time at Observation Point, we sailed to Big Island, where the rocks are pretty much the same as at Elk Island. Among the few fossils obtained here was an interesting specimen of Bryozoa, which has been figured and described by E. O. Ulrich, of the

AMONG THE ROCKS.

Geological Survey of Illinois, Contributions to micro-palæontology, Part II, 1889, as *Diplotrypa Westoni*.

Our 5th camp was made on Deer Island. Here sandstone, green shale and limestone form a ridge of cliffs twenty to sixty feet high—some of the shaly beds are characterized by fucoidal markings which have been, perhaps, twenty feet in length; among other fossils observed here are *Orthoceras*, of large dimensions; *Receptaculites*, which resemble the seed portion of large sunflowers—one or two species of coral, and a few other obscure forms.

Punk Island was the next locality visited. A strong north-west wind drove us on this island, very nearly wrecking our boat. The shore at the east end is white sand, which would make a good glass. This sand is characteristic of several parts of the shores of this lake. There was little to interest us here, and, while waiting for the weather to clear, we amused ourselves with fishing, several "gold eyes" and two fine cat-fish were caught.

From Punk Island we crossed the lake to Big Grindstone Point, on the south shore. Between this and Washow Bay the shore is rugged with shelving cliffs of sandstone and limestone rocks similar to those of the islands already spoken of.

Our lake Winnipeg photographs—which may be seen at the Geological Survey Department, give an excellent idea of a portion of this beautiful lake.

July 31st we crossed Washow Bay to Bull's Head. While our two men went slowly on with boat and equipment, Macoun and I walked many miles along the shore, noting the various exposures of rock, collecting fossils and other objects in natural history.

Low ridges of broken cliffs occupy several miles of the shore of this locality. Many interesting fossils were found here in the vicinity of Dog's Head, where our next camp was made. Here, one evening while wandering alone along the shore I captured with my hands a fine pelican. It appeared to be resting after a long flight, and sleeping; when grabbed by the neck it fought bravely. I shouted, and Macoun ran to my assistance. We wanted this specimen for the museum, or we should have given it its liberty. But how different this fish-eating bird of the wilderness looks in the glass case! The following day (Sunday) Macoun skinned and prepared the pelican while I photographed the Hudson's Bay Post, of Dog's Head—a log cabin occupied by a half-breed Indian, his wife and children. A portion of the building appeared to be a stable for a number of esquimaux and other dogs which were then lying around the buildings. This H. B. Post was the only building seen along the coast.

Near any H. B. Post one is sure to meet with

AMONG THE ROCKS.

Indians; such was the case here, two families, representatives of the Bungo and Swampy Cree tribes. Twelve members of one family were crowded under a poorly constructed bark wigwam. It cost half a pound of tea to get a photo. of this outfit. The other family said they were also Bungo Indians, who I believe are an offshoot of Swampy Crees. These descendants of once brave races of red-skins appear to have discarded, or—perhaps for a few of the white man's luxuries—parted with all the gaudy trappings so dear to their forefathers. With very few exceptions all the Indians seen on the shores of Lake Winnipeg were clothed in the cast-off garments of the white man. One weather-beaten buck considered himself in full dress when he wore a battered stovepipe hat, which was probably made before the present owner was born. Through my boatmen, who spoke this old buck's language, I found he was chief of one of the Swampy Cree tribes, and had lately been married to a young squaw. One of my most valued Indian stone pipes* was obtained from this man for a plug of tobacco.

Many of these old chiefs although very poor are proud and courteous in manner and always find a way of returning, in some small way, any gift or compliment paid them. After trading for the stone pipe and a fine pike just fresh from the lake, I invited the old man to tea with us. We

---

*See Toronto *Mail and Empire*, Feb. 2nd, 1898—Fig. 1.

had canned beef, pancakes, &c., &c., for supper. It was perhaps the feast of the old man's life; he went away with his hand on his stomach and tears of gratitude in his eyes. Just before turning in for the night, we were surprised to see the old man appear at our camp fire with a young squaw—his new wife. Speaking in his own language, and with great courtesy, he presented his bride to me; I called our Indian cook Basque to interpret and found that the old chief presented his wife to the white chief for the remainder of the time he pitched his wigwam at Dog's Head! Through Basque I presented my sincere thanks for the noble chief's kind offer, which I could not accept as I had a white squaw in the big city, and she would not like me to take another into my wigwam. The old man looked hurt and surprised, but a little tea to the fair lady, and a slice of pork to the old man, fixed matters up nicely, and they went away into the darkness of the night.

Continuing our journey along the coast, we reached Kinwow Bay, Aug. 8th. From this beautiful spot we sailed to a small island a few miles distant. This we found to be the most interesting yet struck, so far as fossils are concerned. Thousands of slabs of yellowish white dolomitic limestone have been washed up from the lake, many are much water-worn and bleached by the rain and sun till they look like big pan-

AMONG THE ROCKS.

cakes. These rocks, which are seen in place at low water, dip at a slight angle towards the north. During the few hours we were able to remain on this little island, we collected many fossils new to the Galena limestones of that district; in fact the rocks were so rich in organic remains that, on being requested by the men to name the island, I called it Bonus Island. From this prolific spot of land and rock we crossed to Cat Head. Here a large range of cliffs 60 to 70 feet high occupy the coast. Geologically they are interesting rocks as they contain fossils not found in this formation elsewhere. These cliffs hold a great number of kidney shaped "chalk balls" which give the face of the cliff the appearance of having been bombarded. Some of these forms have a nucleus of flint.

This is the end of our journey towards the head of the lake. Since leaving the mouth of the Red River we have covered about half the length of this magnificent lake, which is about 270 miles long. While returning we encountered several heavy storms and often had to run into some sheltered spot.

Aug. 12th, we were again at Dog's head—this time to find a large number of Bungo or Swampy Cree Indians assembling from all parts. They were congregating to hold their "Dog Feast." This accounted for the constant tum, tum of the Indian drum, which we had heard, wafted over the waters of the lake from distant shores.

The feast was to take place the following day, on the opposite shore of the lake, which here is perhaps only half a mile across—so we decided to stay and witness this feast of the white dog.

After a great deal of formality, in which two pounds of tobacco and a few other traps figured, Chief Sac-a-chew-we-ass allowed the servant of the great white chiefs at Ottawa to point the nose of his camera over the top of the blanket door of their sacred wigwam, which consisted of a long shed-like tent, built of branches of trees, and decorated with any scraps of bright colored ribbons, rags, tin cans, etc., they could beg, borrow or steal. In the centre of this feast-wigwam was a large iron pot, in which the white dog was to be boiled. The occasion of this dog-feast gathering was to initiate young Indian braves, lads who were old enough to be instructed in the art of Indian warfare. When part of the object of our visit was accomplished, through an interpreter, we were politely requested to move off, that the great Dog Feast might proceed. As we pushed our boat from shore, the blanket door of the wigwam was raised to admit twenty or thirty Indian braves (?) who, with rattles made of tomato and other fruit and meat cans containing a few pebbles and fastened to the end of sticks, had been driving away any evil spirits lurking around the ceremonial lodge.

August 17th. We had a splendid run from Goose Island, where we had been weather-bound, to Grassy Narrows. Here we shot several ducks, and Macoun clubbed a pike three feet three inches long. We had little or no provisions left, so these edibles were very acceptable.

We pitched our twenty-first lake camp near Willow Point, and the following day ran our boat into the Red River and made our night camp on a marshy spot of land, the right of which we had to dispute with hundreds of frogs.

At 5.30 the following morning we found a stiff head wind blowing down stream. Rowing against wind and current was out of the question, so the remainder of our boat journey was made by tracking, and at 2 p.m., August 21st, we reached the point at Selkirk we had left on the 19th of July.

Our grub had been poor and scanty for several days, so on reaching the hotel all hands sat down to a jolly dinner, after which Basque and our skipper were paid off; the former proceeded at once to get drunk, while the man of our little craft—who had brought us safely through storms and sunshine—slipped quietly down stream towards his humble home on the north shore of the great lake over which we had safely glided.

Late that night poor Basque was found under

REMINISCENCES

a hedge sleeping off the effects of too much firewater. The money I had paid him—money which would have furnished his family, who dwelt in a wretched shack down the river, with pork and flour enough for the winter—was gone. We spent two or three days collecting specimens and visiting several historical spots on the banks of the Red River. One of these was Fort Garry —renowned for the many attacks its walls had withstood in the fierce warfare with savage Indians. Another interesting spot was the Gun homestead, where we found lodging for a day or two. The Gun family date back to the early settlement of half-breed Indians on the banks of the Red River. Stone implements of war and the chase are often plowed up on the banks of this muddy stream. It was our good fortune to find two fine stone mauls or hammers. Now where the Indian prepared his flint arrow-heads and weapons of war a church stands, and a bell calls both Indian and white man to prayer.

August 30th, we were again in Winnipeg ("Win" muddy and "Nipi" water) the largest prairie city of Manitoba, over 760 feet above the sea; in 1871 a village with a population of 241 souls, which in thirteen years increased to 20,000. Where thirty years ago the red man pitched his bark wigwam, now stone churches, banks, schools, factories, handsome dwellings and stores stand.

AMONG THE ROCKS

At this time, 1884, cheap, hastily erected frame buildings occupied a large portion of the town, prominent among these were many "second-hand" shops, where for a small portion of their original value might be purchased guns, rifles, revolvers, dagger-knives, fishing tackle, cartridge and money belts, together with hundreds of other articles which were to figure in the new life their former owners were to pursue. . . . .

But we must hasten to the locality where our last geological examinations are to be made—Stony Mountain, which place we reached August 30th,—Stony Mountain, that dreaded name to all evildoers in Manitoba, for it is here, a few miles from Winnipeg, on elevated ground, surrounded by fertile prairie land, that the handsome Penitentiary is situated.

In the spacious vestibule of this building were mounted specimens of buffalo and other denizens of the surrounding country; the walls were hung with Indian trophies of war and the chase.

Mr. Bedson, warden of the institution, received us kindly, and after dinner detailed one of his men to show us his fine herd of buffalo, which, with the exception of having only a limited range of their native prairie to roam, are in a state of nature. Two prairie wolves, a badger, a bald headed eagle, and other live stock served to amuse those prisoners who were allowed to leave the jail walls.

REMINISCENCES

The fossils found in the white limestones of Stony Mountain indicate that the geological horizon here is higher than that of East Selkirk and Lake Winnipeg. Our fortunate discovery of *Beatricea ungulata*, a fossil (?) which is characteristic of the Hudson River formation of Anticosti and Rabbit Island L. S., seems to leave no doubt about the formation.

In digging a well to supply the Penitentiary with water a thick bed of purplish colored argillaceous limestone was cut through. This bed is crowded with organic remains—corals, bryozoa, brachiopoda, gasteropoda and other genera are abundantly represented in these rocks, which, but for the excavation made for water, might have remained unknown. Apart from my study of the rocks, bugology occupied a good portion of two nights. Cæsar's ghost ! when I think of those two nights, and yet our hostess expressed no surprise when shown how many boarders occupied one bed.

And now our geological work in Manitoba is over for this season and we are homeward bound.

During our explorations my companion, James Macoun has, whenever an opportunity presented, pursued his favorite study—botany, and returns to Ottawa with an important addition to his father's already large collection of Manitoba plants.

At Barkley—a small way station between Winnipeg and Port Arthur—owing to a big

AMONG THE ROCKS.

wash-out, our train was delayed thirty-six hours. Alex. Mackenzie, M.P., ex-Premier of Canada, and wife were passengers on this train. The old gentleman asked, and I answered, many questions regarding the working of the Geological Survey.

At Port Arthur we found a number of members of the British Association, who, with the Director of the Canadian Geological Survey, were taking a run through a portion of our great North-west.

Another voyage over the waters of Lake Superior, Lake Huron and Georgian Bay to Owen Sound, a railway journey of two hundred miles or so and we were again, on the 6th of September in Ottawa.

CHAPTER X.

EXCAVATIONS AT QUEBEC—EOZOON AGAIN—A COLLECTION OF MAMMALIAN REMAINS—WORK ON THE QUEBEC GROUP—THE BLACK LIMESTONES OF QUEBEC — SUPPOSED COAL DEPOSITS—EVADING THE SCOTT ACT—WITH MR. FLETCHER TO CAPE BRETON — ARISAIG, N.S.—THE COAL MINES OF SYDNEY.

SEPTEMBER 17th. Having heard that some obscure fossils had been found in the rocks then being taken out of the excavations of the Drill Hall on the Cove Fields, Quebec, I lost no time in reaching that locality, and was rewarded by adding to our exceedingly scanty knowledge of the fauna of the Quebec City rocks. Among the interesting collection of Graptolites made were several forms new to this portion of the Quebec group I shall take this opportunity of recording that in all the researches made in these city rocks, not one typical Lévis fossil has been found, such as *Phyllograptus typus*, Hall. Leaving Quebec City rocks, a short time was again spent among the graptolitic shales of Point Lévis, and again, as is always the case in these remarkably prolific rocks, many interesting additions were made to our already large collection of graptolites. Some of these forms show features not known to Prof. Hall when he wrote his important Monograph, Decade II, 1865,

AMONG THE ROCKS.

"Graptolites of the Quebec Group," written for the Geological Survey of Canada.

This was the year of the discovery of that interesting graptolite *Tetragraptus approximatus*.

It seems strange that this species, which is common in some beds of the Lévis shales, should not have been known at the time Hall wrote the above report. My last official journey of 1884 was made Oct. 2nd, to Côte St. Pierre to obtain a new supply of specimens of the supposed fossil Eozoon of which I have already spoken. Another fine collection was made from the locality where Sir J. W. Dawson and I collected in 1874.

Mr. Larique, a farmer, occupant of the small house figured by Dawson in his sketch of the Côte St. Pierre Eozoon locality—"Dawn of Life," p. 22—drove me to Papineauville. Here at the railway station I met two English geologists who had been in search of Eozoon and "the man with the hammer." These gentlemen—Dr. James Murray and G. H. Parke—had been told in Ottawa that I was in the Eozoon locality, so they hunted both Eozoon and the man with the hammer. They discovered the latter just as he was about to start for home, but their precious collection of "Eozoon" did not happen to have a single representative of this distinguished "Laurentian fossil." I gave them several specimens from my collection, and they went away happy.

REMINISCENCES

Although so few geologists now believe in Eozoon being organic, Sir William Dawson, who, by the way, a few weeks ago, celebrated his golden wedding—is still as firm a believer in Eozoon as he was when he wrote his "Dawn of Life." At this time (April '97) I believe his last efforts in behalf of Eozoon are going through the press.

During the winter of 1884-85 many duties connected with the museum and the restoration of specimens for study and exhibition fell to my lot. The first of these was the removal and rearrangement of the ethnological collection from the upper to the middle flat. This being accomplished, the somewhat tedious task of restoring, labelling, etc., the large collection of mammalian fossil bones, collected by Dr. Dawson, Mr. McConnell and myself from the Miocene rocks of the Cypress Hills, N.W.T., was undertaken. There being no osteologist connected with the Canadian Geological Survey, it was decided that this, or a portion of this collection should be placed in the hands of Professor E. D. Cope, of Philadelphia, the renowned professor of comparative anatomy, etc., etc., etc. Having arranged this large collection of Miocene fossils for inspection, the Professor was invited to come to Ottawa and select such specimens as he thought worthy of description and illustration. Prof. Cope came, made his selection, and they were shipped to his rooms in Philadelphia. The result of Professor Cope's studies of this collection can be seen in "Contributions to Canadian Palæontology," Vol. III (Quarto), 1891. Later on I shall speak of Prof. Cope again, but there is an incident I must record here.

Yesterday, April 15th, 1897, while writing the above lines, and while my thoughts were wandering to Prof. Cope's work-rooms, 2102 Pine Street, Philadelphia, where another large collection of fossil bones belonging to the Geological Survey of Canada—chiefly of my collecting—were being studied, I received the Philadelphia "Evening Bulletin" for April 12th, 1897, in which the death of Prof. Cope was announced. I need not say what a shock this was to me. After two weeks' sickness, this great scientist was cut down in the midst of his geological labors. And now shall I have to go to the Quaker City and see to the returning of our specimens?

The remainder of the winter and till June 2nd was spent in over-hauling all the Carboniferous fossils in the museum. Many of these were developed, relabelled, etc., and several weeks at microscopic work completed the indoor duties for 1884.

The short time allotted me for field work during the summer of 1885 was spent on the rocks of the "Quebec Group." The want of fossils to determine the geological horizon of the Quebec City, and a large portion of the Eastern

Township rocks, was the cause of my researches this summer. As the following notes will show, Dr. Selwyn, director of the Geological Survey, was much interested in this "Quebec Group" work, and joined me several times for the purpose of studying the stratigraphical structure of this formation. Arriving at Quebec June 23rd, I found our chief awaiting me at the St. Louis Hotel. Together we carefully went over the rocks of the "Cove Fields" (already mentioned). I was then left alone to examine the rocks of the city, and here I must record one of the most interesting fossil discoveries made in the shales and limestones of this place.

It seems—but why, no one knows—till this time to have been taken for granted that with the exception of two or three species of graptolites found last year in the "Cove Field" beds, a few fragments of crinoidal columns and plates, and two species of Bryozoa, discovered by the writer during Logan's connection with the Survey, that these city rocks were destitute of fossil remains. The result of these discoveries and those of the last three or four subsequent years has proved that the limestones, shales and conglomerates of Quebec City hold many interesting organic remains; in fact, a large fauna of both graptolites and other genera, among which those found this year—1885—deserve special mention. I refer chiefly to the discovery in the rocks near the back of St. John's St., or Montcalm Market. Here I found the black shales and limestones to contain numerous small bivalve shells belonging to the *Lingula* family, also a great number of minute Crustaceans belonging to the Trilobite family. Specimens of the latter fossils were forwarded to Mr. Ford, of Troy, N.Y., who was then engaged in the study of the trilobites of the primordial formation. Ford's interesting report on this collection can be found in "Transactions N. Y. Academy of Science," vol. 7, 1897.

On June 25th, Dr. Selwyn again joined me in the study of the black limestones of the city, more particularly those of the quarry near to St. Patrick's Street, where this rock has been largely quarried for making hydraulic cement. They are the characteristic limestones of the city, and a portion of the north-west end of the island of Orleans, and contain one or two species of graptolites which occur abundantly in the Hudson River formation. On June 27th, the well-known surgeon, Dr. Ahern, of Quebec, accompanied me to Point Lévis, where the previous year he had discovered, near Fort No. 2, fine specimens of the typical bivalve shell, *Lingula Quebecensis*. A few good specimens were obtained, but I had to leave further researches for some other time, which at present has not come. Dr. Ahern has frequently done the Geological Survey good service, and to this day remains one of my most esteemed friends.

On July 1st, Dr. Selwyn again joined me, and together we examined some miles of the rocks on the south side of the island of Orleans. Two or three days were then spent between the east and west end of the island, on the north shore.

Our present visit afforded us a good opportunity of visiting the so-called coal deposits, which were then being so much talked of. We found the locality, but as usual these "coal beds" turned out to be a deposit of black shale, owing its blackness, in part, to being very much slickensided—caused by two beds of rock rubbing together—and to thin veins of Anthraxolite, a metamorphosed or hardened bitumen, a mineral found in thin seams in the various formations of the "Quebec Group." I think I have mentioned before that the rocks of the island of Orleans are many thousands of feet below the coal beds of Nova Scotia, and were deposited many thousands of years before any coal deposits we know of.

The following incident relates to the first outcry I heard of coal on the island of Orleans.

Somewhere near the year 1868, one or more gentlemen waited on Sir William Logan and informed him that a deposit of coal had been found in the rocks of the island of Orleans, and requested that he would accompany them and make a report on this discovery. "Fiddle-de-

AMONG THE ROCKS.

dee ;" said Sir William ; " the rocks of the island of Orleans are thousands of feet below the carboniferous rocks, and we have no coal in Canada." "But, Sir," replied one of these men taking some lumps of coal from his pocket, "here are specimens from our coal beds." " Yes," replied Sir William, " these are good samples of Newcastle coal." The men were persistent, however, that they had a valuable coal deposit, and at last Sir William consented to visit these mines. A careful investigation of the locality resulted in the discovery that many years ago a blacksmith's forge had stood on this spot, and later on a rock slide had buried the blacksmith's forge and the coal which constituted this valuable coal mine.

In 1846, Sir William was called on to inspect some " coal deposits " in the vicinity of Baie St. Paul, below Quebec. Here with the exception of a patch of Trenton rocks the formation is Laurentian. The upshot of Sir William's researches here resulted in the discovery that imported coal had been carefully " packed " to deceive the inhabitants.

Other similar incidents might be related with regard to the discovery of coal in the Province of Quebec, but the above will serve to show that at present no true coal has been found in beds lower than the Carboniferous formation, which formation is not represented there so far as we know. Our nearest coal deposits are in Nova

Scotia. Continuing our journey eastwards from the " coal mine " partly by buggy and partly on foot along the margin of the island, we were soon opposite the magnificent Falls of Montmorency, which are said to be nearly a hundred feet higher than those of Niagara. Here the foamy waters leap 250 feet over the face of rugged rocks. A view of the falls from Orleans island is worth going a long distance to see ; but the grandeur of this mighty body of water rushing from its summit to the valley below can be seen to its best advantage only at the foot of the fall, where if one is so inclined he can in a few moments get drenched to the skin in the icy spray which rises like a cloud from where the fleecy waters bury themselves beneath the waters of the river. Some years after the time of which I am writing, while making a typical collection of the rocks of the falls, I experienced what it was to get a bath from this icy spray. Perhaps it was the exertion of climbing the 400 steps to the top of the cliff that counteracted the chill received.

Taking to our buggy again, we continued east along a road dotted with farms owned by French habitants, descendants of the early French settlers of the island. The numerous wild plum trees and vines of this island perhaps account for Jacques Cartier calling it the " Isle of Bacchus."

AMONG THE ROCKS.

Arriving at St. Famille—a thoroughly French village "as old as the hills"—on Saturday afternoon, the doctor decided we had better remain there till Sunday, so our driver, a French Canadian, interviewed the only people who kept boarders, and after a considerable amount of talk we were at last quartered till the following morning. After a sumptuous dinner of pork and beans my good chief—who never lost an hour working time unless his strength gave out—proposed that we should descend the banks to the shore and proceed towards the east end of the island. The shades of night were falling when after a four hours' geological tramp we returned to our French Canadian boarding-house, where no one seemed to speak English. We were both a little played out and I proposed taking a little stimulant, if it could be got in a village where the "Scott Act" was in full force. Hunting up our landlord, and taking him to one side, I placed my hand on my stomach and said: "Monsieur, mon ami et moi bien fatigués, nous-avons mal-dau cœur, avez-nous du quoi, du whiskey ou du brandy." The old man gazed in a frightened way around the room, then with one finger up to his nose to enjoin silence locked the front door. Dr. S. and I were then taken into a back room, the door of which was also locked; then the old man unlocked a quaint old-fashioned corner cupboard and produced a bottle of gin; well, we had our drink, but while taking it I felt like a thief and was glad when the doors were unlocked and open once more.

At 8 o'clock the following morning (Sunday), we were in our buggy again and crossing to the south side of the island. The geological section presented along the south shore of this pretty island, affords an excellent opportunity of studying a number of the characteristic rocks of the "Quebec group."

Towards evening we reached the west end of the island and again took up our quarters for the night in our friend Lizotte's comfortable hotel—long since destroyed by fire, but restored again on a much more extensive plan by the enterprising owner who, like our friend Gee, of whom I shall speak later on, has sheltered and fed many members of the geological staff, and to whom so many of the improvements of the west end of this pretty island were due.

July 6th. We crossed over to St. Joseph de Lévis where the government graving dock was in course of construction. It is a structure 484 feet long and 100 feet wide. This great work has long been completed, and I have there often walked under the keels of our large ocean steamships. In excavating for this great dry-dock many interesting beds of the Lévis rocks were exposed. One large block of conglomerate taken out we thought worth having cut into slabs for the museum, and it was shipped for that purpose.

From St Joseph we took the I.C.R. track to Point Lévis, stopping every little way to examine rock-cuttings, some of which yielded many fossils new to the "Quebec Group." We returned to Ottawa that evening, July 6th.

Part of the months of July and August were spent in the Eastern Townships examining the rock exposures near Richmond on the St. Francis River and Hereford, where I crossed the boundary line into the State of New Hampshire.

The first indication I got that I was in Uncle Sam's domains was a cemetery with most of the graves decorated with small American flags, flowers and children's toys. It looked strange to us to see all these little Yankee flags stuck into the graves, but since then I have seen many American burying-grounds, and find the people of the United States pay much attention to the decoration of graves.

The absence of fossils in the great stretch of rocks examined rendered this my last official journey of 1885 monotonous; still it was pleasant to travel along the banks of the picturesque St. Francis River, almost every foot of which was known to my driver, who was our old friend Gee, whose house at Melbourne had for months, sheltered our greatly respected late chief Sir William Logan, Richardson, Webster, myself and

REMINISCENCES

other members of the Geological Survey staff. Poor Gee! Incidents relating to his " historical house " crowd thickly on my memory. Many pages might be written about the " Gee homestead," where Logan spent his last working days. Gee was never more happy than when driving his renowned guest, Sir William, to and from his field work in distant localities. All of the old members of the staff of those days—excepting Dr. Bell, who is still actively engaged in field work for the Survey, and myself (at my own request I was superannuated in 1894)—are gone to rest. Logan, Murray, Billings, Richardson, Barlow, Hunt, and many others, who, though not members of the staff, took an active part in the Survey work. Last of these (to the present time, 1897) I must mention our old friend Gee, of whom I have been speaking. Poor Gee, who fed us on the fat of the land (?) and delighted in piloting us to some new discovery he had made of serpentine, slate or other minerals. I believe this geological excursion of 1885 was the last Gee ever assisted in. But even though all these late members of the staff have passed away, the geological work—as the reports of the Survey will show—still goes on, being diligently pursued by Dr. R. W. Ells and other members of the staff.

As usual, the winter months of 1885-86 were occupied in office and museum work. During this season in conjunction with Mr. (now Dr.) Ami, many improvements were made in classifying, arranging and labelling specimens, preparing samples of gold and Eozoon for the Colonial and Indian Exhibition.

My geological excursions of 1886 commenced early in July, when I joined Mr. Hugh Fletcher, B.A., one of the Survey's most energetic field geologists. The geological work accomplished by Mr. Fletcher in Nova Scotia and Cape Breton will remain forever a monument to his energy and perseverance. It was to assist this gentleman in researches for fossils to determine by palæontological evidence the geological horizon of certain rocks, that I joined him. To be with Mr. Fletcher while he is pursuing his field work is a geological feast. This, with the beautiful country through which we travelled, was refreshing indeed after spending weeks on the non-fossiliferous rocks of Lake Huron and the Georgian Bay. With Mr. Fletcher I drove and walked through a good portion of Antigonish, gleaning from the rocks much evidence of their past history. About ten days were spent examining the various rock exposures in the vicinity of beautiful Lochaber Lake, near the shore of which we boarded with a charming family, who, after the day's duties were over, gathered around the piano or organ and sang the good old melodies of long ago. A game of euchre with the old folks, or a sail on the lake usually ended the days—days all too short for both work and pleasure.

From Lochaber Mr. Fletcher and I journeyed to Arisaig, N. S., which place we reached July 20th. Since my first visit to this locality in 1869 many changes had taken place. The tedious buggy drive of 30 miles from New Glasgow is now accomplished in a short time by rail. Changes, too, had occurred in my old friend Joseph McDonald's homestead; marriages, deaths, departures to fields more promising than this rough sea coast and the scanty livelihood to be got from these rocky lands. But " The sea rolled on as it rolled before," and here, as we travel under the sea-splashed cliffs, we see no change and decay. I have spoken of the geology of the Arisaig rocks before, and shall only mention here that many new fossil forms were added to our already large collection.

Aug. 8th Mr. J A. Robert—one of Mr Fletcher's assistants—joined me, and together we proceeded to Cape Breton to examine some of the Cambro Silurian rocks of that region. Railway, stage and steamboat carried us to North Sydney, where I had an interesting chat with Mr. Richard Brown, manager of the Sydney coal mines. To this gentleman the Survey is indebted for many beautiful Carboniferous coal plants.

Leaving Sydney we drove to George River, Bras d'or Bay, where, in the vicinity of Mr.

AMONG THE ROCKS.

McFee's farm, we discovered beds of limestone holding numerous fossils belonging to the genera *Obolella*, *Lingula*, *Hyolithes*, &c., forms, some of which have been described by Mr. Matthew, of St. John, N. B.

After wandering along the shores of Bras d'or Bay—which were strewn with thousands of jellyfish cast up by the waves—over verdure-clad mountains, along the shores of Escasoni Lake, where we fought the rocks, and caused them to yield up many precious geological relics, and where we boarded at some of the picturesque farm houses in the vicinity of Christmas Island settlement, where some of the people end most of their sentences with " whatever," we wended our way to Riversdale station, on the Sydney branch line of the Intercolonial Ry., a short distance from Truro. Here we found lodgings at the Post Office, kept at one of the few houses in the immediate locality of the station, where the arrival and departure of trains formed the chief excitement of the place.

We spent five quiet, restful days with the homely old lady and her daughter, who carried the mail-bag between the station and the house.

In this quiet old-fashioned home, where the table in the best room was strewn with goody-goody books, to my surprise I found a charming novel, " Broken Toys," by—I forget the author —written many years ago, evidently by one who understood the ways of children well, for it gives a beautiful description of a little girl who tried to bring up a large family of dolls in the way they should go, but with all her love and care some of them went sadly wrong; one lost an eye, another a leg, then Mary Jane became so thoroughly demoralized that, as an example to her other sisters, she had to be buried and gotten out of the way for ever and ever.

The rocks in the vicinity of Riversdale Ry. station represent a portion of the Carboniferous or Devonian system. The railway, a little east of the depot, cuts through a thickness of about 115 feet of argillites of various colors, beds of sandstone and layers partly made up of concretions. Some of these beds are prolific in the bivalve shell *Naiadites* (Dawson), while other strata hold numerous fragments of fossil plants, chiefly *Cordaites* leaves. *Calamites*, *Neuropteris* and many other fossils of the Carboniferous or Devonian period occur in these interesting rock sections. This is a locality where many days might be spent profitably collecting.

August 31st, I passed through Point Lévis and spent a few hours on the graptolite beds, adding one new species to our list.

Oct. 7th, I was again at our famous Eozoon locality at Côte St. Pierre, where, thirteen years ago with Sir William (then Principal) Dawson, we studied in place the supposed Laurentian fossil.

AMONG THE ROCKS.

Again this year, 1886, with Glen Gordon, of Ottawa, I was entertained in the little farm house made historical in that interesting book—"The Dawn of Life." But I have spoken several times about this supposed Laurentian fossil, and shall only say here that, although cart loads of pieces of this supposed ancient coral reef have been carried away, hundreds of tons weight still remain for those interested in *Eozoon Canadense*.

While writing the above I have received a letter, and several photographs, from Mr. Fletcher, of the Survey, dated May 20th, 1897, in which he says: "A crowd of us, including Dr. Selwyn and Adams, went to Côte St. Pierre last Saturday to find *Eozoon*," which, he adds, some think "will die a natural death."

During the winter months of 1886-87 I was occupied as usual with microscopic and other work connected with the ethnological and palæontological branches of the Survey and museum. It gives me pleasure to record here the advice and kindly aid I always received from those officers of the Survey who were more learned than myself in some matters I had to deal with. My thanks are specially due to Mr. Whiteaves, Mr. Fletcher and Dr. Ami, for kind assistance in anything pertaining to literary work.

CHAPTER XI.

EXPLORATIONS ON THE SOUTH SHORE OF THE ST. LAWRENCE—FISHING FOR COD—A PLEASANT STAY AT MR. RICHARDSON'S—A JOURNEY TO GRIMSBY—A TRIP TO THE NORTH-WEST PLAINS—MAPLE CREEK—SWIFT CURRENT—BIG JAW COULEE—A PRIZE PIG—THE BAD LANDS—LETHBRIDGE—WE LAUNCH A BOAT ON THE RED DEER RIVER BUT AFTERWARDS GIVE UP OUR PLAN—THE JOGGINS COAST ONCE MORE.

DURING the winter of 1886, doubts regarding certain portions of the geological structure on the south shore of the St. Lawrence, below Lévis, existed in the mind of the Director of the Survey and he decided to visit this section of the "Quebec Group" the following summer, and that I should accompany him.

Accordingly on the 14th of June, 1887, I joined Dr. Selwyn in Quebec and together we proceeded by caleche to Beaumont, the small village next below St. Joseph de Lévis. Here we bagged some lunch and started down the shore in search of various outcrops recorded by Mr. Richardson. We had a most interesting tramp, and many facts regarding the relation of the south shore rocks to those opposite on Orleans Island were obtained. The following day we examined the exposures in the vicinity of St. Michel de Bellechasse. Along this section of the coast, many fine examples of anticlinal and synclinal foldings are seen. Miles

AMONG THE ROCKS.

of red, green and gray argillites, quartzites, limestone, conglomerates, and other rocks tilted at various angles, form a rugged coast. There is considerable evidence that the rocks in this locality are lower than the Lévis proper.—Since the date of which I write, the geology of this shore has been worked out and mapped by Dr. Ells of the Survey.

Continuing along the shore towards Lévis we came on what Dr. Selwyn supposed to be the fault between the Lower and Upper Lévis. The former horizon as regarded by Selwyn extended past the snow shed on the Intercolonial Railway. But here in the shed, fossils came to our aid, and for the first time since the "Quebec Group" controversy typical Lévis fossils were found, among which was the bivalve shell *Lingula Quebecensis*, thus proving that we had passed the Lower Lévis beds, and were then on the Upper Lévis formation.

On our return to Point Lévis Dr. Selwyn left me, and I spent several days alone collecting fossils from the cliffs at Lévis and from the shore rocks in the vicinity of Cap Rouge on the Quebec side of the river.

On the 7th of July, I again left Ottawa for Quebec to join Dr. Selwyn, who was to accompany and spend a short time with me and Mr. Lambe—the artist of the Survey, who was to be my assistant—on the rocks in the vicinity of Gaspé and Cape Rosier.

REMINISCENCES

In a former visit to Cape Rosier I had discovered compound graptolites which indicated that the escarpments in the vicinity of the lighthouse were older than they had hitherto been supposed to be, and it was chiefly to see these graptolitic beds that the chief of the Survey accompanied me. Going by rail to Dalhousie, and by steamer Admiral to Gaspé was a delightful trip. The Government S. S. La Canadenne which was about to leave Gaspé Basin with lighthouse supplies, carried us to Grand Grève where through the kindness of Dr. Wakeham, commander, we were landed by one of the steamer's boats. A cart carried us from Grand Grève to Cape Rosier lighthouse where we were kindly received and accommodated with board and lodging. Although it was the 11th of July, the weather was cold, and we were glad to sit by a wood fire and listen to the waves as they dashed against the rugged cliffs a short distance below our bedroom window. The following morning our worthy chief, anxious to see the rocks in which I had a few years before discovered the compound graptolites, hurried us off to the shore below the lighthouse. For a short time I failed to find my graptolite beds, and think the Doctor doubted my veracity. However in due time I pointed to some thin bands of blackish gray limestone, perhaps thirty feet up the cliff, and said, " you are a good climber, Sir—will you see what those black beds contain?" Up went the doctor, and in a few moments, holding a piece of limestone in his hand, and with a beaming face called out: "You are right, Weston, here are your compound forms." While the doctor threw down pieces of the rock Mr. Lambe and I made a selection of the best fossils, some of which are now in the hands of Prof. Lapworth, of Birmingham, England, for identification. The remainder of the day was spent along the coast east of the lighthouse. Two days later the chief left us and returned to Gaspé to join his daughter, who had accompanied us from Lévis to that place.

During the third night of our stay at Cape Rosier lighthouse, while the wind roared and the angry waves lashed the shore, a wee wail from a new-born babe was blended with the roar of the sea, and we were called on to drink the health of the little stranger who had come to help swell the French-Canadian population.

Mr. Lambe and I left our snug retreat in the lighthouse July 19th, and proceeded by cart along the coast road, making our geological observations and collecting fossils wherever we could find them.

We reached L'Anse au Griffon July 20th, where bold cliffs of argillite and other rocks occupy the coast. The rocks of North Bluff at Griffon Cove are prolific in graptolites, many of which are characteristic Hudson River forms.

REMINISCENCES

In a hay cart we pursued our way along the shore road, which, as my companion Lambe observed, "was just rough enough." At Cape Magdalen we spent a very pleasant day or two, boarding as usual in fishermen's homes where each member of the family vied with the others in trying to make us comfortable.

Black slates containing graptolites of the Hudson River formation are well displayed in the vicinity of the lighthouse, and the sand stones of this vicinity hold a few Brachiopods. From Cape Magdalen lighthouse we examined the coast rocks to Mont Louis and Gros Mâle. Graptolites of the Hudson River series occur in abundance in some of the beds of black shale.

The scenery of this section of the river is very fine; in fact all along the shore of the grand stream one finds an endless variety of scenery, and a constant change in the arrangement of the rock formations. The pleasure, however, of a journey such as the one I am speaking of, is much greater when with camp and provisions—as in my first journey up this coast—as then one is independent of the (very often) poor, but cheerfully given accommodation in fishermen's homes.

We reached the Marsouin River July 29th. This is another very interesting geological locality. Its black shales are abundantly stocked with graptolites, presenting a rich feast for the palæontologist.

AMONG THE ROCKS.

A considerable amount of fishing is done at the various places near where we found lodging for the night. If at some of these stations one has time to spend half a day cod fishing with one or two of the sturdy fishermen of this coast, he will find food for the mind as well as food for the body. Generally I am not a successful angler, but standing one evening on a block of stone a few paces from the shore, I cast my fishing line, with a hook baited with pork, into the dark waters. In a few moments I felt a vigorous tug at my line and hauling in I found a good sized cod on my hook. In a shallow stream which empties into the river, small trout were abundant, but as they feed on the decaying refuse of the cod fish, prepared here for market, I did not disturb them.

On the first of August we reached Ste. Anne des Monts, where the St. Anne River—one of the finest salmon streams in Canada—empties into the St. Lawrence River. I have already spoken of this river and my journey up it to the Shickshock Mountains, in 1878.

The coast here presents many interesting features to the geologist and palæontologist. Here an extensive series of dark colored slates and shales occur, in which we find—three miles or so above the mouth of the St. Anne River—typical graptolites of the "Quebec Group," the most prolific forms are the well known *Phyllograptus typus* and *P. angustifolius*.

REMINISCENCES

Above Ste. Anne, black, green, red and other colored slates and shales with massive beds of conglomerate are seen for miles up the coast. The fossils found in the limestone conglomerates leave no doubt as to their geological horizon, viz., " Bic conglomerates " of the " Quebec Group."

Following this varied strata we reached Ruisseau A'sem, Aug. 4th, where we were hospitably received by Mr. John Richardson, son of James Richardson of whom I have frequently spoken in these memoirs. We found Mr. Richardson busily engaged in his saw-mill, where thousands of cords of birch wood had that summer been stowed, and were then being cut into long strips, tied up in bundles and shipped to Scotland and other places for making spools. At this time a large barque was being loaded with bundles of this product—birch cut from the extensive forests near the shores of the St. Lawrence River—some of which would no doubt in course of time be returned to Canada in the form of cotton spools.

The weather was delightfully fine during our stay at Ruisseau and we made long tramps along the rugged shore, returning at sunset to find the busy operations of the day suspended, and the inmates of Mr. Richardson's refined seaside home gathered around a cheerful wood fire. After a pleasant evening, in which music and cards figured, and while the mighty waves dashed against the great cliffs just below, each member of the household read a verse from a chapter in the Bible, after which a prayer followed in which the "strangers within our gates" were not forgotten, and the good Lord was asked to guide the footsteps of those who sought sermons in stones.

We left this delightful retreat Aug. 5th, and followed the coast to Grande Matane where we were kindly received and entertained at the house of Mr. and Mrs. James Russell—son-in-law and daughter of the late Mr. James Richardson.

Following the coast to Little Matane we passed over large exposures of black slates, some of which are rich in graptolites, especially in two or more species of *Dictyonema*—fossil Polyzoa, or Hydrozoa, which resemble on a small scale the coral sea fan *Gorgonial.*

Having finished as far as time would permit, our coast examinations, Mr. Lambe and I proceeded by rail to Point Lévis, where after a few hours on the ever interesting rocks of that locality, we returned to Ottawa.

The excellent water color sketches of some of the coast rocks made by Mr Lambe, on this his first geological journey, together with the series of photographs taken by myself in 1878, serve to remind us of pleasant days—days spent in trying to unravel the complicated structure of the Quebec rocks, pleasant days for those who, as

REMINISCENCES

John Richardson said in his prayers, "sought sermons in stones."

On the 14th of September I was again in Grimsby, Ont., that pretty little village near the shores of Lake Erie, and a few miles from Niagara Falls. Here again I met my old friend Mr. Johnson Pettite who was ever ready to lay aside his books and "bugs" and accompany me in a fossil hunt. But I have spoken of Grimsby and our friend Pettite before, so will only mention here that again we had a profitable time among the (in places) highly fossiliferous rocks of the Medina and Niagara formations. Besides securing many good specimens belonging to the order *Echinoidea*, or sea urchins, we collected many other fossils of interest, and the large slab of *Arthrophycus Harlandi* which adorns the wall of the geological museum. It is supposed to be a fucoid or sea plant—*arthron* a joint, *phykos*, a sea plant.

September 17th I again settled down to my usual winter duties in the Survey building. Many new specimens were added to the museum and much good work done by each member of the staff—work done not simply for our daily bread and butter, but for the love and interest each member of the Survey took in the welfare of the institution.

During the early summer of 1888, I was again preparing for another long journey across the

AMONG THE ROCKS.

North-west plains. Our former collection of mammalian and reptilian fossil bones had proved so interesting that it was thought advisable to visit the Cypress Hills district again, and then proceed to Red Deer River, Alberta, where in 1887 Mr. J. B. Tyrrell of the Survey had obtained a large portion of a Dinosaurian skull and other reptilian fossil bones.

On June 28th, provided with camp equipment I left Ottawa and proceeded by Canadian Pacific Railway, about 1600 miles, to Maple Creek which place I reached July 2nd. Here I was again indebted to Horace Greely, of Powers and Bros. general store, for fitting me out with waggon, buck-board, horses, two half-breed Indians and provisions.

We left Maple Creek July 4th, and struck out for the head waters of Swift Current. On reaching a prairie farm a few miles distant one of the men repented his contract and refused to go further with us. I was just about to return to the Creek to try and get another teamster when a typically dressed cowboy stepped up and said: "What's the matter with me going with you uncle; I don't like boastin', but there aint many places around these parts Joe Lowry and Billy don't know, and for that matter Billy and me kin take you any wheres between here and California."

I found Billy was a French Canadian pony

that had been the constant companion of his master Joe in all his wanderings from the Pacific coast to the Atlantic, and back to our present locality. I liked Lowry and we soon proceeded on our journey—Jim our teamster with the waggon, Lowry with the buck-board, I on horseback, and Billy following as close to his master as he could get. We reached the head waters of Swift Current July 10th, and camped in the main valley—called Frenchman—from which many coulées branch. It was in one of these coulées, on our present explorations, that the largest portion of the skull of an extinct Rhinoceros was found.

Big Jaw coulée, through which a beautiful stream of clear cold water flows and in which bush and green pastures form a delightful retreat for the antelope and a few other wild animals who still roam the Cypress Hills and valleys, besides yielding to us a large portion of the extinct Rhinoceros skull, has been the home of ferocious flesh-eating animals—animals nothing but the bones of which have ever been seen by the eyes of man, who appeared on this planet ages after these creatures had ceased to exist.

Lowry became greatly interested in my researches and determined that he too would add something of value to our collection. I was making the geological section which appears on a previous page, when Joe, who was far up the

AMONG THE ROCKS.

other side of the coulée shouted "Uncle—" the only name Lowry ever gave me—"I've found the big bone you wanted." I found the specimen to be a well preserved femur or thigh bone of our extinct Rhinoceros, or as Prof. Cope calls it—*Menodus angustigenis*.

It would require too much space here to record all the interesting incidents of our journey over the plains, and through the valleys of the Cypress Hill district.

A few more words however—which by permission I copy from my field-note book—and for this year we must leave our Miocene rocks with their wonderful records of a silent world—silent except for the roar of savage beasts as they sought their prey among the peaceful animals of those past ages.

Big Jaw Coulee N. W. T.
Sunday July 15th, 1888.

"8 a.m.—Bright clear morning—Bar. 26.1 Ther. 54. strong S. E. wind. 9 p.m. Scout (Lowry) and I have ridden about ten miles over this locality and have visited one of my bone coulées—I think the one we named Cold Water coulee, as besides having a stream of water through it, it had several springs near the mouth. This is the head waters of Swift Current, about five miles north of our camp or where the Pie Pod trail enters the main coulée (Frenchman). The day is closing in very cold—I have to keep on my over-coat. My two cow-boys are merry to-night, owing to the prospect of starting for civilization to-morrow."

Two lunch and one night camp—which was made at Bear Creek on the East Post trail, and we had again crossed the Cypress Hills and are

REMINISCENCES

at Maple Creek busy shipping specimens, settling accounts, &c.

Barring the little accidents almost always met with in crossing the North-west plains the journey, to one who loves nature, is full of pleasure, especially if one has men like those I have had on this trip—who when the day's work was done sat around the camp fire and told of their wanderings in distant lands; of their cow-punching incidents, feats in lassoing wild steers, &c. I was inclined to doubt some of the tales Lowry told of his lassoing exploits, but while at "The Creek" I had an opportunity of seeing that he was no boaster in the art of lassoing. While we were waiting for our train we joined a prairie picnic which was going on near the "town." Among a number of prizes offered for various feats, was a pig which was to be given to the first man, Indian or white, who could lasso Mr. Pig at a distance of fifty yards. The pig was dumped out of a box on the prairie flat; a sharp prod in the hind quarters started him off at full gallop, but in a moment six lassos were thrown, in another moment a wild shout from hundreds of Indians, cow-boys and the motley group which always composes a prairie picnic—for Lowry had his lasso round the pig's neck, thus winning the prize, which I believe he sold for $5.00.

On July 22nd, I was again at Lethbridge, and the following day (Sunday) with a team of horses,

AMONG THE ROCKS.

waggon, saddle-horse, two men—one of whom was a typical North-west mule driver—and provisions for a few days, we started for Milk River Ridge.

I have stated before that the collection of fossils Dr. Dawson, Mr. Tyrrell, my son George and I made in 1883 were all lost by the sinking of the ship on which they were freighted. It was with the hope of replacing some of these specimens that the present journey was made.

Soon after leaving the settled portions in the vicinity of Lethbridge bad luck seemed to follow me. Travelling the bleached clays and sands of the "bad lands" of the North-West Territories under favorable circumstances has not many redeeming features, but in a blazing sun with one's horses almost mad with the stings of that great pest of the North-West, the horse fly, and one's self swarmed with mosquitoes, to have to listen constantly to the blasphemous language of two of the lowest types of prairie mule-drivers, takes away much of the pleasure of a geological journey.

Many miles away from civilization we woke at sunrise to find our horses gone. Jim stood for a few moments, using language not fit for these or any other pages, then without breakfast started to track our horses The sun had set when he returned with our three horses and two extra ones he had picked up on the plains. It was during Jim's absence that the fine large bivalve fossil shells which are to be seen in a wall case in the Survey Museum, were discovered; other fossils of much interest were found in the natural excavations of these barren lands. Here was the tooth or rib of some extinct reptile, and there the cast of a delicate shell whose pearly beauty had long perished, and here again the leaf of a plant which received its name many thousands of years after its verdant life had decayed. These and many other relics of past ages were gathered from the rocks of the Milk River Ridge.

I was not sorry to hurry away from these dreary, though highly interesting bad lands of the Milk River district, and rid myself of two men who were pessimists of the first water. We reached Lethbridge July 26th, where men and travelling equipment were at once returned, and the hearts of an Indian and his bride, who arrived at that moment both seated astride an Indian pony, were made happy with the remains of our provision larder.

During the winter of 1887, I had, through correspondence with the Rev. Leo. Geatz of the Red Deer Crossing (about 100 miles from Calgary), arranged to have a large flat bottomed boat built in which to descend to and below the confluence of the Red Deer and South Saskatchewan rivers. I left Lethbridge July 27th, and arrived at Cal-

AMONG THE ROCKS.

gary two days later, where a short time was spent on the rocks of the Bow River, where the Canadian Pacific cuts through high banks of gravel, sandstone, and shales. The two latter series of strata hold many fragments of fossil plants, but the most prolific rocks in fossil flora are those of the "Hog's Back," one mile down stream from the railway depot. Here in a fine display of the Belly River sandstones, capped with drift deposits which together form a bluff of perhaps 200 feet in height, large fossil leaves of exogen plants, associated with smaller leaves and flowering plants were found. Some of these interesting fossils have been described by Sir J. W. Dawson in Proceedings of the Royal Society of Canada. Those figured, are from drawings made by my friend L. M. Lambe, the artist of the Geological Survey. These are relics of an epoch ages prior to the time when man appeared and shaped his weapons of war and the chase out of some of the stone found here, and when no human eye had seen the mighty sea, which has left tracings of its winds and waves in these stony records. Early on Thursday morning, August 2nd, perched on the top of a large lumbering stage coach I started for the Red Deer crossing, which, as I said before, is about 100 miles north of Calgary. The journey to one not acquainted with prairie life is one of much interest. Our first night was spent at a prairie farm where, after a good supper of pork and potatoes, with six other men, rolled in buffalo robes, we slept on the floor.

At five the following morning the horn sounded for breakfast, and at 7 o'clock we were again on our road. The heavy rains of the two previous weeks had made much of the trail almost impassable, and many times the horses were knee deep in black peaty muck. Our stage broke down several times. Only an experienced driver, which we had, could have got through these sloughs. One of the interesting features of a journey to Edmonton from Calgary is the meeting of hundreds of "freighters," all half-breed Indians, men who from infancy have toiled along the prairie trails with their heavily laden carts.

We reached the small village of the Red Deer Crossing Aug. 5th. Here I found the boat I had expected to be ready, only just commenced, so I pitched my tent and waited. The Red Deer Crossing village consisted of two general stores; one run by the son of the Rev. Leo Geatz, who has a residence and farm about a mile distant, and one, during the absence of the owner, kept by Mr. W. B. Vennor, brother of the late H. G. Vennor, late of the Geological Survey of Canada. The latter gentleman will be remembered by many as at one time holding the important (?) position of Canadian weather prophet. These

AMONG THE ROCKS.

two stores and a few scattered houses form the welcome resting places of many weary freighters before and after crossing one of the most rapid streams of Alberta.

I failed to find a man who had been more than ten miles down the Red Deer River from the Crossing, in the summer, and was glad to get two half breeds who were willing to leave civilization for an uncertain length of time, and venture down a part of a river unknown to them or any of their associates. But at last these two men were engaged, and while waiting the completion of our boat, which was being built by half-breed Indians from pine cut on the bank of the river a few miles below the crossing, and sawn for that purpose, I had several rambles with my friend Vennor—who years ago had left his position in the Montreal Bank to wander in his "adopted country"—who pointed out many places of geological interest and caused me to assist him in making a collection of the wild flowers of that district which appeared to be, in variety and abundance, more beautiful than in any other portion of the North-West I had been over.

Monday, Aug. 13th, our boat was launched, and after taking dinner in the parsonage and farm-house of the Rev. Mr. Geatz and his charming family we three men, with camp equipment and three months' provisions, embarked in our rudely constructed boat, and drifted down the swift current, watched by all the members of the Geatz homestead, who made their adieus and wished us God speed. I think from their anxious faces they had doubts as to whether our boat would carry us far, before some calamity overtook us. If they did think that, they were right, for in a short time I found that we were incapable of guiding our heavy-laden boat through the swift waters, in fact my men seemed to know nothing about navigating rapid streams. Part of our first night was spent in trying to patch up our boat, and the following morning we succeeded in reaching Mr. McKenzie's farm, eight miles below the crossing, and the last settlement on the banks of the Red Deer River, before entering an unsettled and little known portion of that land.

Mr. McKenzie, a noble specimen of the halfbreeds of that district, was greatly surprised to find us attempting to descend the river equipped as we were, and strongly advised me to abandon the journey, promising if I did, to have another boat built by the following spring and to accompany me himself. This I decided to do, and in a short time boat and part of our provisions, camp equipment, &c. was stored, my men settled with and soon I was on my way back by stage to Calgary. It seemed hard after so much trouble to abandon my long looked for trip, but as McKenzie said we could not afford to lose our lives for a little geological work.

I reached Calgary Aug. 18th, and Irvin, or Ross Coulée, the following day—I have already spoken of this interesting locality—and after spending a few days gleaning new facts regarding the geology of this district, returned to Ottawa, which place I reached August 30th.

On Sept. 9th, I was again on the Joggins Coast, Nova Scotia, and with my old friend Mr. James Devine, a coal miner—who to assist me for a short time, had left his pick and shovel far down in the bowels of the earth—was again searching the shore rocks for new geological information. It must indeed be a poor student in geology and palæontology who cannot find abundant material for study from the magnificent cliffs of the Joggins Coast, which are constantly crumbling away and exposing to view erect sections of fossil trees, trees in the base of which small reptiles—some of the first representatives of vertebrate animals of this world—crept for safety, trees whose branches overhung shallow waters in which some of the first fishes known by these fossil remains, sported; waters on the shores of which grew delicate and noble ferns, grasses and other plants; representatives of our early endogen and exogen flora.

The view of the Carboniferous rocks as represented on the Joggins coast is one of great interest; its massive beds of sandstones intersected by bands of coal, beds of bituminous shale holding millions of bivalve-looking fossils, which some writers think belong to the Crustacea family; beds of limestone in which scales of extinct fishes are mingled with delicate jaws of minute reptiles. But to unravel and read from these stony records an epoch when the great coal deposits of Nova Scotia were formed has taken years of patient work, to which the writer of these notes is proud to have been privileged to add his mite.

A stage journey of about three hours brought me again to Macan where, after dinner and the shipping of my specimens—one of which was the large section of a fossil tree, now exhibited in one of the upright cases of the geological musem —I boarded the west bound train on the Intercolonial Railway, and two days later was again in Ottawa.

CHAPTER XII.

SUPPOSED FOSSIL TREES—A VISIT OF INVESTIGATION TO KINGSTON—I START FOR THE RED DEER RIVER—MR. MCKENZIE JOINS ME AT CALGARY—EXTENSIVE COAL SEAMS—FINE DINOSAUR REMAINS—CANADA HAS DONE COMPARATIVELY LITTLE TOWARDS RESTORING THESE DENIZENS OF A SILENT WORLD—DANGEROUS QUICKSANDS—WE CELEBRATE DOMINION DAY—A LONG RIVER TRIP—ANOTHER TRIP TO THE CYPRESS HILLS—UNPLEASANT EXPERIENCES—ON THE BORDERS OF LAKE SUPERIOR.

DURING this year, 1888, much discussion prevailed concerning the discovery of supposed fossil trees, many examples of which are seen in the Cambrian (Potsdam) formation on the banks of the Rideau Canal, Kingston, Ont. The attention of the director of the Canadian Geological Survey was drawn to these singular forms; he visited the locality and caused a section of one measuring four feet in diameter to be sent to the museum in Ottawa. (Notes on concretionary structure, by the writer; Transactions of the Nova Scotian Institution of Science, Series 2, Vol. 1.)

On the director's return I was requested to visit the locality, and on the 16th of October, accompanied by Mr. Topley, the Government photographer, we started, and on arriving at

REMINISCENCES

Kingston found several people much interested in the discovery of these so called "fossil trees." In a shop window we saw a section of one labelled "Section of a stone tree." Hiring a team we drove through the historical city of Kingston. Our twelve mile drive to the celebrated quarry where these so-called fossil trees occur was a pleasant one.

We found quite a number of these cylindrical forms, some examples would weigh many tons; some of the weathered sections show irregular concentric rings which resemble somewhat the lines of growth in exogen plants. No fossil trees have yet been discovered in the Cambrian formation to which these deposits belong, and the conclusion arrived at was that these tree-like bodies are of concretionary structure, formed probably in geyser cavities. The "potato-stones" found in these sandstones are nodules, formed of the same material which composes these rocks.

The photographs taken at this interesting locality are the property of the Geological Survey and can be seen at that institution. This completed my geological journeys for the summer. The winter of 1888, was fully occupied in office and museum work ; work to me always interesting and instructive, and oftener than not a labor of love.

On Saturday, June 1st, I again started for the Red Deer River. Dr. G. M. Dawson. now Di-

TREE-LIKE CONCRETIONS AT KINGSTON, ONT.

rector of the Geological Survey, left at the same time for British Columbia, and we travelled together as far as Calgary, which place we reached June 6th.

Comfortably seated in one of the luxurious Pullman cars of the Canadian Pacific Railway I often thought of our first journey over a portion of this railway, when we shared part of a freight car with three horses, dismembered waggons, bales of hay, etc., and took our meals in tents and rude buildings by the track-side.

At this time, during the construction of the C.P.R. through the districts of Saskatchewan, Assiniboine, and Alberta, the red man of the plains, together with their dark-skinned squaws and children, all attired in bright colored blankets, beaded leggings and moccasins, feathers and paint, assembled at various points of the track to watch the doings of the white man. As the status of the Indian is estimated by the number of ponies he has, these "cayuses" always accompany their owners, who flocked to watch the iron horse plough its way through their late hunting grounds, and to pick up a few pence from the travelling tenderfoot in exchange for polished buffalo horns and other trinkets made in their "spare time."

Although only a dozen years or so have passed since our first journey over this country, it is remarkable what a change has taken place in the appearance of many of these children of the plains. The blanket, by most, has been discarded for the cast-off clothing of the white man, and many bucks and squaws out of sheer necessity are working alongside the white man. Still there are many elderly aborigines who scorn to labour, and will, so long as life and memory last adhere to the pursuits of their forefathers.

June 6th.—Mr. McKenzie—a fine example of the half-breed Indians of the Canadian North-West Territory—joined me at Calgary. He had made the journey of a hundred miles or so with waggon and horses from his farm on the banks of the Red Deer River, to meet and convey me to my starting point down this—so far as vertebrate palæontology goes—one of the most interesting rivers of the North-West.

After another short examination of the fossiliferous rocks of Calgary, from which several fine specimens of fossil leaves—one 12 inches in length—were collected, and a supply of provisions laid in for our journey, we started for the McKenzie farm. I have spoken of my previous journey by stage, from Calgary to the Red Deer Village. Our present trip by waggon was much less interesting, and occupied a day longer. Part of it was made on the Edmonton stage road, and part over the plains, till we reached the banks of the Red Deer River, opposite the McKenzie farm—eight miles or so below the Red Deer Village—

AMONG THE ROCKS.

which place we reached at 10 p.m. July 13th. Our horses had long scented their stables and were very impatient to cross the river, then a very rapid stream which had risen a foot since " Mac." left, and it was a question whether we would not have to go to the ferry, eight miles up stream. Mac. gave me the reins while he examined his sounding marks on the banks of the river. The night was cold, and in the distance we heard the barking of numerous prairie wolves or Coyotes (*Canis latrans*); I had anything but a soft job to hold in our horses till Mac. returned and said we would cross on his lower ford. Our camp equipment was arranged to keep it from the water which Mac. expected would cover the floor of the waggon, then he took the reins and with the cheering words, " keep your eyes closed and don't move," allowed the horses to plunge into the stream. For a moment I felt the waggon sway with the current and the water covered its floor, but in another minute or two our horses gained a footing on the opposite shore, and with a last brave pull ascended the steep bank, and in a few moments I was comfortably seated before a large fire in the hospitable home of one of the early pioneers of the Red Deer River district.

June 15th was spent at the mouth of the Blind Man River, eight and a half miles up stream, where a fine geological section of the Edmonton series is seen. One of the beds of rock here holds beautifully preserved fossil plants; leaves which have been pressed in nature's stony-book—pressed so closely that every vein is plainly seen. Our small but valuable collection of fossil plants from this locality, together with photographs showing the confluence of the Blind Man and Red Deer rivers, form an interesting souvenir of this picturesque spot where we saw a graceful deer come to the water's edge, drink and then glide away into the bush.

My friend McKenzie, who was versed in several Indian languages, and had spent many years among the aborigines of the Canadian North-West, gave me many interesting accounts of the early settlers and explorers of the Alberta and other districts; which reminded me of having read that among the earliest explorations of which any record remains, we find the name of M. Bigot, the French Intendant, who after his succession to the intendancy in 1748, sent out explorers to spy the land in various parts of unsettled Canada, hoping, it is presumed, to add to his already luxurious and libidinous ways of living.

In Alberta, as in almost all sections of the North-West Territories, many of the beautiful names given by the Cree and other Indians to lakes, rivers and mountains have been changed by the white man, but between themselves I found the Indians used the old Indian appellations, and even the half-breed seemed to resent the re-naming of rivers, localities, etc., a thing too much done by young explorers of the present day.

The following morning, June 17th, after a successful day's work at the confluence of the Blind Man and Red Deer rivers, the McKenzie family, typical representatives of the half-breed Indians of the Canadian North-West were assembled and photographed, then our two boats were launched and provisioned, and at 2 p.m. we waved adieu to our friends of the last habitation we should see for many days, and drifted swiftly down a portion of a river unknown in the summer to any of the settlers I had met with in Red Deer village. This is probably due to the swiftness of the current, which in places is so rapid that it is almost impossible to track a boat up stream again. The waters of the Red Deer flow from one of the snow-clad interior ranges of the Rocky Mountains, where, after coursing through the foot-hills it is a smooth stream of clear blue water two hundred feet wide.

Soon after leaving the McKenzie farm we reached the "cañon" in which are high, steep and scarped banks. Here we found the water rushing between large Laurentian boulders, rendering navigation dangerous; to me the scene was very fascinating. Once the bow of my boat was driven on a projecting boulder, and in an instant the stern swung around and our flat bottomed craft was heading in that fashion down stream. I heaved a sigh of relief when a few moments later we swept into smooth water. In a short time we reached the great, yet (1889) undisturbed coal seams of this district. Here an almost vertical section of seams of coal rises fifty feet above the water's edge and as far as can be seen below the surface of the river. In 1858, when Dr. Hector navigated this stream, large exposures of this coal were seen to be burning, and now the large outcrops of red and yellow burnt coaly matter lend a picturesque aspect to this vicinity.

Bold escarpments several hundred feet high occupy portions of the river-sides for many miles, and from the table-land, sloping valleys, thickly wooded in places, form lovely retreats for the numerous wild animals that roam the sheltered recesses. Along the shore of the river and muddy flats we noticed the tracks of grizzly and smaller bears, coyotes or prairie wolves, woodchucks or ground-hogs. the prairie dog, and other rodents. I must not forget to mention the beaver, individuals of which tribe were seen in several places along the margin of the river. One fine fellow was busily engaged in the construction of a dam, but disappeared the moment he caught sight of us. Outside a deserted

AMONG THE ROCKS.

hunter's shack—near the confluence of the Red Deer and South Saskatchewan rivers, I saw several skeletons of these animals. It is a pity that this valuable rodent, which together with the maple leaf, forms our Canadian emblem, should be becoming extinct on many of our rivers and streams, where a few years ago they were numerous. The rancid smelling oil secreted near the root of the tail of the beaver furnished the castor oil of the early medical man. I remember reading, I think in some book of fables, of these animals, when being pursued, biting out these oil-glands and casting them before the hunters. We all know that the castor oil now used is made from a bean.

As our boats glide silently past numerous small thickly wooded islands and occasionally verdant valleys, we constantly hear the twittering and sweet song of birds, among which is the soft cooing of the wood-dove. The high sandy buttes, where little vegetation, excepting wild sage grows, is the home of the bald-headed and other eagles. On the ragged projecting portions of these escarpments, out of reach of the prowling coyotes, these noble birds build their nests of twigs and buffalo bones; I found in one nest a tin tea spoon.

In the twilight of the evening the hoarse cry of an owl is frequently heard, and the king-fisher is seen to dart from his river-side perch and swoop down on an unlucky "gold-eye" who has ventured too near the surface of the water.

The banks of the Red Deer River are a favorite resort for wild geese; the proverbial stupidity of these birds is very noticeable. Many times after our night camp, we saw a few hundred yards away, a flock of geese feeding on short mossy grass which grows on the muddy shore of the river. They remained feeding till our boat came within gun shot, then there was a hasty, partly running and partly flying, not into the bush or fields, but straight ahead of our boats, stopping to feed when we camped for lunch, or landed to examine some favorable spot for fossils, and then on again till night. Sunday June 23rd, we were fairly in the bad-land district of the Red Deer River. The great sandy buttes and escarpments of this part of Alberta extend for many miles, sometimes coming close to the river side, and in other places lying back, leaving room for grassy and sandy flats. These great exposures belong to the Laramie or upper Cretaceous formation. In this geological horizon, far below the Miocene Tertiary in which the remains of an extinct rhinoceros, already referred to, were found—is a series of rocks of special interest to the osteologist as they contain the remains of great Dinosaurs, those terrible flesh-eating reptiles which inhabited certain portions of our North-west, at a time when what is now prairie land and rivers was occupied by great lagoons and jungles. On one of the great sand and shelving sandstone buttes, a hundred or more feet above the river, and in close proximity to several eagles' nests, we came on the remains of one of these huge Dinosaurs (*deinos*, terrible; *saura*, lizard)—the skull and other bones of which now occupy a prominent position in our Geological Survey Museum. In the sandstone beds that contain these remains I first saw the two lower jaws; the right ramus covers a portion of the left, concealing the teeth, which are seen in the other jaw. A close examination showed that the rami lay upon the roof or palate of the cranium. Only the roots of the teeth of the upper jaws remained, but the teeth of one of the lower maxillaries is almost perfect—large, curved, and beautifully serrated at the edges. There were also portions of limb-bones and claws— dreadful claws—the sight of which carried one back into past ages.

Besides these remains we found slabs of sandstone covered with ripple marks and rain-prints showing us that in those remote ages—before any human eye had gazed on the wonderful works of our Creator—the wind blew and the rain fell as at the present time. Overlying this Dinosaurian grave—the bones in which have been since determined by Prof. Cope to be *Lælaps incrassatus*— were slabs of rock holding beautifully preserved leaves of exogen plants, and grasses.

REMINISCENCES

In the geological section of this earth's crust, these fierce Dinosaurians occupied a period in the animal kingdom millions of years after the Devonian and Carboniferous age, when the first fish and smaller reptiles appeared on life's stage.

When one contemplates the vast ages which have rolled away since the days when these great kangaroo-like saurians disappeared, never again—we may presume—to be repeated on this planet, and then think that the Laramie rocks of which we are speaking, only occupy one leaf of nature's great stony records, our minds are bewildered among the immensities of the past. Scientists tell us that while it takes or has taken 730 years to deposit one foot of some of our sedimentary rocks, other formations have taken 6,800 years to deposit one foot. Thus, remarks one writer, the period of time required to build up 100,000 feet of sedimentary rock has varied according to locality, from 37,000,000 to 680,000,000 of years.

Leaving the high buttes and desert-like sandy flats which have yielded so many important geological specimens—food for the mind—and continuing down the beautiful but sometimes treacherous Red Deer River, we pass through verdant valleys, in places 150 feet below the prairie level. A flock of wild geese keeps ahead of our boat and a beaver slides down the bank into its sheltered dam, a little further on as our boat glides swiftly over a long stretch of rapids we pass a

AMONG THE ROCKS.

hungry-looking coyote, who is evidently looking for a breakfast of goose, and towards evening find ourselves among a series of similar rocks to the Laramie, but lower in our geological section; they belong to that part of the Canadian Cretaceous known as the Belly River Series. As in the Laramie formation, we find these sandy buttes and escarpments strewn with fossil bones, chiefly fragmentary—here several vertebræ twice the size of those of the buffalo; here the distal end of a femur or thigh bone, which must have belonged to a larger animal than that great elephant "Jumbo"—whose skeleton now adorns one of the American museums. These are the bones of Dinosaur reptiles who lived prior to those already spoken of. But besides these great carnivorous beasts there lived in those remote ages, huge herbivorous Dinosaurs, one of which has been named by Prof. March *Stegosaurus ungulatus*. It is remarkable for a series of armour plates on its back and for its small head, compared with the size of its thigh bone. Scattered on these sandy flats are the remains of turtles, scales of Ganoid and other fishes—creatures who existed ages before the Mastodon and Mammoth appeared, and at the time when the lignite coal of our North-west Territories was being deposited. The restorations and writings of Cope and March furnish us with a great amount of information about these extinct animals, especially about those remains found in the United States, where a large sum of money has been spent in pursuing this most interesting branch of fossil osteology. Compared with the researches of the United States, however, comparatively little has been done in Canada towards the collecting and restoring of these wonderful denizens of a silent world—silent except for the roar of these wild beasts and the occasional raging of the tempest. While wandering alone over one of these desolate bone-strewn spots, and wondering whether any human foot had ever trodden the sandy ground, I stumbled over a human skull. It was bleached and weathered and is probably the skull of an Indian who lost his life in one of the fierce battles waged between the Bloods and Crees. This skull is remarkable inasmuch as the atlas or first cervical vertebra is firmly ankylosed to the occipital or base of the skull, which peculiar defect must have deprived Mr. Indian of the pleasure of turning his head either to the right or left. This skull is now in the ethnological room of the Geological Survey Museum, Ottawa.

The patches of quicksand met with in parts of the river bottom and on the flats near the stream, often caused us much trouble. In some parts of the river where the water spreads out considerably our boat, though not drawing more than six inches of water, frequently stuck fast in the

AMONG THE ROCKS.

sands which are continually drifting with the current, and will in a short time completely surround one's boat unless it is immediately got out into deeper water, not always an easy task to accomplish as our boat-pole sunk its full length without finding bottom. On one occasion I was much struck with the danger of quicksands. I had been rambling alone over some of the high sandy buttes in Range 21, Tp. 33, and when returning to our tents stepped on what I thought hard sand, in a dried up brook-course, only two yards or so wide, and in a moment down I went, up to my waistcoat pockets. Almost before I had time to think I had stretched out my arms and grabbed a small bush which grew on the opposite bank, and with some difficulty pulled myself out, to find I was covered up to my waist with ice-cold slimy sandy mud. I scraped part of this mud off with my sheath-knife, and then hastened to my tent which was hidden behind a clump of cottonwood trees. On turning this obstruction to a straight path I almost stumbled over two coyotes who had evidently been trying to get into our provision boxes. Although these sneaky wolves are very cowardly, it is not pleasant to be in close proximity to them, for when hungry they will sometimes attack anything.

When my men returned to camp I got McKenzie to sound the place which had almost engulfed me, and when no bottom was found with our fifteen foot boat-pole, Mac looked at me and said : " Well we should have knowed where you had went to, for your hat would have been left."

The old saying that birds of a feather flock together, holds good regarding the distribution of fossil bones over the bad lands of the Red Deer. After leaving the interesting locality where the Dinosaurian jaws were found, I walked over many sandy buttes and flats without finding anything of interest. The portion of river we had been travelling for the past few days, passed through valleys 600 or more feet deep. Small but well wooded islands formed sheltered nooks for our night camps. Often the surface of the water fairly bubbled as hundreds of " gold eyes " snapped at flies.

In this vicinity, while one's boat glides swiftly with the current and one is congratulating himself that he is making probably ten miles an hour, he finds his boat stuck fast in quicksand. Then it is get out if you can, with your boat pole, but if it is not quicksand one can drag the boat into deeper water.

Sunday, June 30th, was spent in exploring some of the buttes in the vicinity of Range 21, Tp. 30. Here in places the buttes and escarpments came close up to the river. In one of the ranges seams of coal are seen ; further on a burnt butte exhibits in a marked manner the stratification of various coaly layers or beds, some of

AMONG THE ROCKS.

which are of a bright Indian red while others are of various yellow tints. The river scenes in this vicinity are charming, a flock of wild geese keeps ahead of our boats, a large eagle hovers above us, and the constant twitter of birds adds to the pleasure of our journey. The following morning was bright and beautiful. My men had been looking forward to this day for some time, for it was Dominion Day, and we were to celebrate it by tapping our only bottle of brandy which I had guarded diligently in case of getting a rattlesnake bite. Hitherto our "nips" had been made of a decoction of pain-killer, sugar and hot water, a splendid drink when one is "chilled to the bone." We were early in our boats, and while drifting with the current, passed several beaver dams, shot a wild goose, and had climbed a high butte by 11.30 Here we hoisted a long pole we had brought up, which with a large red pocket-handkerchief served as a flag staff. Having christened the elevation "Dominion Butte," fired a salute, sung "God Save the Queen," drunk to all absent friends and relations, we descended to our boats which we reached just in time to find that they were being inspected by several Indians. These were the first people of any sort we had seen since leaving the McKenzie farm. We found that these red skins were on their way to a far distant lake, where, through some of their tribe, they had heard that there were buffalo. Mac got uneasy and motioned me to get in the boat, which I did at once, and as we shot down a stretch of rapids he said: "It's a d—— good job, while we were away, they did not take boats, provisions and everything they could lay their hands on."

For two or three days after leaving Dominion Butte we struggled with sand bars (for the river was very low), mosquitoes, head winds, etc. But as an offset to these little troubles we had rousing camp-fires and jolly suppers of fish (gold eyes), which were—so our cook expressed it—"too quick to take the bait, and so spoilt the sport." In this section of the country, one must be a good sleeper, indeed, not to be continually disturbed with prowling rodents, the barking of coyotes and the croupy cry of the big grey owl.

But morning comes, and with a fair wind and rudely constructed sail at the stern of our boat, which seems to dance over the rippled waters, we forget the troubles of yesterday and those ahead of us.

July 5th, we had reached Dead Lodge Cañon, Range 13, Tp. 24. In this vicinity the river cuts through a fertile valley from 400 to 600 feet deep. Here nature had used her scooping shovel to an enormous extent, for between the prairie level and the river sandy buttes interstratified with bands of sandstone form pyramid-like structures. Here to my delight I found on the bleached sands numerous fossil bones, and further search revealed the fact that this at one time had been the home or rendezvous of both flesh-eating and herbivorous extinct animals, among which figured our old friend, *Lælaps incrusatus*, Cope, for here in the cut side of a high butte sticks out the proxhemal end of a femur which we know to be that of a Dinosaur. Three hours were spent digging out this bone, and three pairs of hands carefully lifted our precious specimen to put it in the rude box we had made from part of the upper floor of our boat, when to our surprise the thing crumbled into a thousand fragments. Many of the bones found in these formations are solid stone, the bony pores having been filled with silicious matter. In this case, however, only a thin outside crust remained, while the interior was a decayed mass which crumbled into dust when exposed to the atmosphere.

In these great sand banks must be buried thousands of tons of fossil bones, for as the weathering away goes on, these relics of distant ages weather out and remain on the sandy flats or roll to the foot of the banks and cliffs. These bones belonged to animals that existed in that wonderful epoch called the Reptilian age, turtles six feet long held their own with the great kangaroo-shaped carnivorous reptiles who with their curved knife-like teeth would have been formidable foes even to the great Mammoth and Mastodon who did not appear till ages after the great Dinosaurs "had laid their bones down," as one of the men expressed it. It was with great reluctance I left this interesting bone locality, but provisions were getting scarce, and we still had to explore many miles of a country unknown excepting for the valuable map of this river, compiled by Messrs. McConnell and Tyrrell of the Canadian Geological Survey.

Sunday July 7th, we waved adieu to what is probably the most important field in Canada, so far as bones of extinct animals is concerned. It would take years to glean from these great sandy buttes, flats and cliffs even a part of the information they hold relating to that period when the Cretaceous rocks of Canada were laid down—a time contemporaneous to that when crocodilian reptiles swarmed the banks of the river, now part of Philadelphia, and the lagoons of Pennyslvania. Continuing our journey we still pass elevations varying in height to perhaps 600 feet, but the Unio and other bivalve shells they contain tell us that they are a little higher in the geological horizon.

The river is still low, and though our boats only draw seven or eight inches of water we have continually to cross and re-cross, keeping them in the small channels made by the current.

AMONG THE ROCKS.

Laurentian boulders of almost every variety are scattered over the table-land, the buttes, and the bed of the river.

Sunday 14th, we reached the confluence of the Red Deer and South Saskatchewan rivers. In this vicinity there are some fine flats well timbered with cotton-wood, black birch, elder, willow and other woods. During the past few days we have seen many flocks of wild geese, a few ducks, several coyotes, and Mac had no trouble in supplying our camp with fish, but with the exception of an occasional black bass it was always "gold eyes."

Near the junction of the two rivers we came on two "shacks." These were the first signs of civilization we had seen since seeing the small band of Indians at the foot of Dominion Butte, so we went ashore (as Mac said) to see how the cat jumped. Several sharp raps on the first door brought no response, and as the door was not locked we entered and found probably all the worldly possessions of a North-West trapper. A candle stuck in a black bottle adorned a rough home-made table, several guns and a rifle hung on the wall, there was a cupboard in one corner of the room, but " when we went to the cupboard we found it was bare." At the back of the shanty we found the festering remains of coyotes, beavers, skunks and other animals. With a blazing sun and the thermometer registering 90° in the shade the stench of this decayed matter was so great that we only remained to tack our cards on the door.

Here at the junction of the two rivers the Saskatchewan takes a northerly course and at a distance of about 70 miles, in a straight line—but probably about double that distance by water—passes through Medicine Hat (of which I have already spoken) where one of the fine iron bridges of the Canadian Pacific Railway spans the stream. But from the mouth of the Red Deer we continue westward.

Monday July 15th, at 8 a.m. the thermometer registered 80° in the shade. For the past few days we had been drifting with the current down the Saskatchewan where in many places the water is both swift and deep—quite a relief after the shallows of the Red Deer. Mac tells me that these rivers, after heavy rains, rise six inches in one night, so the next persons to follow our tracks in search of geological treasures, may be more fortunate, and not even know half the difficulties one experiences when the water is unusually low.

The distance from the mouth of the Red Deer River to the Battleford and Swift Current crossing is—not counting the many windings of the river—about 100 miles. A great part of this distance is uninteresting to the geologist, as the river is flanked on both sides with mud banks and the low valleys are covered with vegetation; the sportsman, however, can find plenty of work for his gun and dogs.

Leaving our Belly River rocks—so named because they are largely developed on a river of that name—which have yielded us so many interesting fossil bones, we—20 miles or so before reaching the Saskatchewan Landing—come to a series of strata made up of sand, clays, muds, sandstone and limestones. They represent the upper part of the Canadian Cretaceous known as the Pierre and Fox Hill Group. We hastily pass over these few remaining miles, and at noon, July 19th, reach the Battleford and Swift Current crossing. Here we find a small settlement of half-breed Indians; many of these hardy sons and daughters of toil have been occupied from almost childhood in freighting over the great plains of our North-West. Some of these freighters can tell wonderful tales of the fierce struggles between various tribes of Indians who waged war with one another in this vicinity where the old worn-out freighter now cultivates his little garden, while the stronger members of his family continue their freighting occupation.

We pitched our tent for the night on the shore near a "shack" or log cabin which was occupied by one of the North-West mounted police officers. Here, in exchange for one of our boats, we obtained sugar and pork which we were much in need of.

The following morning having arrangèd with a freighter to meet us with a cart, in four days, we drifted down stream about ten miles and pitched our tent on a verdant plateau opposite Swift Current coulée.

I have already mentioned that the geological formation of this section of the country is the upper part of the Cretaceous. This formation here, is characterized by high cut banks, buttes and flats. These deposits of various colored clays, sands, etc., in places, hold concretions or nodules of iron-stone, in which are found beautifully preserved fossil bivalve and other shells ranging in size from a quarter of an inch to a foot or more in diameter, the latter being the large bivalve *Inoceramus*, which must have held meat enough for a hungry man's dinner—had he existed in those days.

Among the great variety of fossils found in the upper Cretaceous—which represents the upper part of the Mesozoic rocks of Canada so wonderfully developed in this locality,—we find several species of Ammonites,—representatives of the Cephalopoda family—so called because they have some resemblance to the horns of the Egyptian God, Jupiter Ammon. In the rocks and clays of this locality we frequently find portions and sometimes whole specimens of these forms, some of which measure two feet in diameter, and on the outer shell show all the beautiful opalescent colors as when in ages ago these "shells of the ocean" floated on silent seas —seas no human eyes had ever gazed upon. Over these clays are also scattered the fossil remains of animals whose form and character we shall probably never know, for the osteologist who can build representations of some of the extinct vertebrate animals out of a limb bone fails to recognize some of the specimens from this locality.

Now I have come to our last night's camp on the shores of the Saskatchewan River. Long after darkness had closed in, my two faithful half-breeds and I sat beside a roaring camp-fire and talked of days when the aborigines of the North-West pitched their wigwams on these shores and engaged in all the cruelties of Indian warfare. But hark!—the crackling of our camp fire does not prevent the half Indian instinct of Joe hearing an unusual sound, and in a moment he seizes my rifle and creeps through the bush followed by Mac with a gun. For a time all is quiet, then a rifle shot is heard, and in a short time the two men return to camp, but without any trophy of the chase.

Presently I turn in and sleep soundly till towards morning, when I awake to find a brilliant reflection of our camp fire on the walls of my tent, and to hear Mac singing:—

'—Prowling—beasts—about—Thy—way,
Stones—Thy—pillow, earth—Thy bed.'

In his young days, Mac had attended the Mission School on the banks of the Red River near Winnipeg.

We found that Joe's rifle shot wounded something, for there were blood stains on the grass. While at breakfast the sight of an Indian freighter wending his way over the rough undulating country on the other side of the river, was a signal for us to strike camp and cross the river which just then, owing to the recent rains we had had, was exceedingly swift and, as after all heavy rains, very muddy. In a short time we had crossed the stream and had dragged our boat above high-water mark, there to leave our good but roughly constructed and now much battered craft, which had carried us safely over hundreds of miles of the two winding rivers we had hastily examined for the remains of a world uninhabited by man.

Our traps were soon packed in the freighter's cart, and after a journey of ten miles over hills and through valleys we again reached the ferry crossing.

I spent the night in the police shack while the men camped outside. Early next morning, with our dozen or more boxes of fossils weighing

AMONG THE ROCKS.

about a thousand pounds snugly packed in freighter's carts, we left the banks of the Saskatchewan River, and following the Battleford trail arrived at Swift Current at sunset. The distance, although only thirty miles, presents many interesting features incident to prairie travel, especially when one makes the journey, as we did, with an old-time freighter (and his numerous family) who had spent most of his life on this trail and could point out spots where, as our old guide said, brave Indians buried their tomahawks in each other's skulls. But now here all is peace, and the "iron horse" of that wonderful Canadian Pacific Railway snorts its way through the quiet village of Swift Current, where a few Indians still remain, loath to quit their buffalo hunting grounds of a few years ago, loath also to assist the white man, yet forced to make a scanty living from him. The following morning I bade good-bye to my two genial companions of many days and nights spent "near but yet so far" from the haunts of the white man.

July 26th. I was again at Maple Creek preparing for another trip to Cypress Hills. I have already spoken of the verdant hills and valleys of this bright spot in our North-west Territories, and recorded a few of the fossil remains found in the Miocene Tertiary rocks exposed in the coulées and hills of this locality, so shall hastily pass over the present journey to these interesting rocks—rocks which have thrown much light on the fossil fauna of the North-west.

Having hired a wagon, three horses, one a saddle horse, and two men, and with a month's provisions, we left Maple Creek July 27th, and followed the trail James Macoun and I took in 1884.

Between Maple Creek and our present destination I notice many changes have taken place during the past five years; land then covered with prairie grass is now under cultivation. In other parts ranches appear, and the typical cow-boy herds his cattle; further on, where the trail descends into a deep valley, we pass the bleaching bones of a horse—" rode to death while tracking an Indian horse-thief " says one of my men. Still further on in a valley we come to four Indian graves; stones that cover the recently turned earth have not prevented the hungry coyote from digging up portions of the buried bodies, for fragments of limb bones and a piece of a skull lie near by. Some distance from this locality we came on another Indian grave, but this is in a tree—a rudely constructed wooden box rests on two limbs of the tree, out of the reach of prowling beasts. We looked into this box and found the remains of an Indian woman, who in life was no doubt loved and considered beautiful, for beside her had been placed two skilfully worked bead bags, each containing pigments for painting the face; a string of glass beads and a few brass ornaments completed the toilet articles which were to adorn this dusky lady of the forest and plain when she entered the happy hunting grounds of her forefathers.

The great numbers of buffalo bones that lay bleaching on the plains and in the valleys are now all gone—gone to refine sugar and fertilize other lands—gathered up by the same Indians who a few years ago only gathered the hides and a little flesh of the now almost extinct noble bison which roamed these prairie lands in vast herds. Only, in many miles, one skull was seen, and this served as a finger-post, for on it was roughly painted—" water 100 yards north."

We pitched our tents near the head waters of Swift Current where a large coulée entered the main valley of this district.

About a week had been spent collecting fossils and gaining new geological information relating to our Miocene deposits, when to my surprise I found a deadly enmity had sprung up between my two men, one of whom I had days before discovered was a deserter from the American Army—a man of the very lowest type of the scum of the earth. For several days the sound of this man's blasphemous tongue had grated on my nerves so much that I was almost unfit for work. This together with the loss of one week's provisions, "stolen by our Yankee soldier" said my other man, who was not a bad fellow, decided me to return to Maple Creek as soon as possible. Quietly loading my rifle while talking to these men I ordered them to load up at once as I intended to sleep at the police camp that night. It was thirty miles away and we had nine hours to do it before sunset. With rifle in hand I mounted my horse and waited patiently till all was ready for the start. Our Yankee was particularly sullen and I feared he would not go; but he knew full well that if he deserted me he could not appear in Maple Creek again, so at last we started. Following the waggon enabled me to keep a sharp eye on Sam, for that was our Yankee's first name. Our road lay through valleys and over high buttes and was an exceedingly tough journey.

Some one has said that the vilest of men have some redeeming traits, so I found it with this man. Large herds of half wild cattle—unmolested excepting during the branding seasons—wander up and down the valleys of the Cypress Hills district. In one of these valleys we found a young steer mired in a creek which crossed the valley. Without uttering a word our Yankee friend (?) jumped out of the waggon and motioned the driver to stop, then taking two picket-ropes from the waggon beckoned him to assist. In a few minutes the ropes were fastened around the horns and body of the poor beast, then the horses were taken from the waggon and the ropes attached to the whipple-pole, and the almost exhausted steer dragged to dry land, probably to die, for it was too feeble to stand when we moved on. During our long journey, this was the one redeeming act of our Yankee outlaw.

Half an hour or so later in a narrow part of the valley we came on a large herd of cattle, creatures who for a few years of their lives range as much at will as the buffalo who once darkened these valleys with their dusky and bulky forms. We were close on these cattle, and for a moment I wondered what these cowboys would do to get a way through this moving mass of animals. But when we were within a few yards of the head of the column, which appeared to be headed by several very fierce looking bulls, one of the men turned towards me and motioned me to put spurs to my horse and ascend the banks of the valley then each man seized a picket-rope, and with great shouts swung them around their heads, at the same time urging the horses on. Out of danger I watched the slow progress of our waggon through this great drove of cattle, which probably numbered two thousand. I breathed a sigh of relief when at last the team passed safely—with the exception of a nasty wound one of the horses got.

REMINISCENCES

Our way now lay over a very rough portion of the country, so much so that in descending some of the hills all four wheels of the waggon had to be firmly lashed together. But we reached the police station (shacks, or log cabins) safely, where I was hospitably received and sheltered for the night, while the men camped outside. To the sergeant in charge I related the trouble I had had with my men and that serious trouble might happen before I got them back. The following morning, Sunday, I found a splendidly-mounted police officer had been detailed to escort me "into town," he had received instructions not to leave my side till we were in Maple Creek. We rode ahead of the waggon, but twice the police officer had to return, and presenting his revolver order the driver to "get a gait on." We reached our lunch camping-ground, where the buffalo's skull was, with the notice; "water 100 yards north"; then after giving our horses an hour's rest continued and reached Pie-pod's Creek and at 6 o'clock our camping ground for the night. Here we found a large encampment of Indians, aborigines who prefer a roaming life and often hungry stomachs rather than remain on the Government Reserves with daily rations of food. This (Bear Creek) is a favourite camping ground, and just then there appeared to be an unusual gathering of red-skins, for some special purpose. We would have preferred spending the night elsewhere than in such close proximity to these Indian camps, but as this was the only place we could get water, for some miles, we had no option.

After supper our horses were safely picketed a few yards from our tents, and at 11 o'clock our police escort took off his boots and coat, spread his blanket and was soon asleep by my side. Although tired after being ten hours in the saddle and much worried, as several annoying incidents (due to my men) had occurred during the day, I did not doze off till near midnight. Then, with the exception of a faint light in two or three of the numerous tepees, and a few prowling dogs, all was quiet, but it seemed I had scarcely closed my eyes when a tremendous discharge of fire-arms, accompanied with savage yelping, startled me, and at the same moment the police officer sprung to his feet, and with revolver in hand rushed from the tent; returning in a few moments, he said "without my boots, my horse takes me for one of those b———y red skins and will not let me go near him. I'm not afraid that they will do us any harm, but I am afraid they might play some trick on my horse." This man's magnificent horse—although the night was unusually dark—recognized his master at once, with his boots on, and allowed him to lead him to our tent and picket him close to us.

The Indians continued firing and shouting

their war-whoops till early morning, when all at once they quieted down to gambling, a fact indicated by the constant beating of several "tom-toms." But soon after they were up and preparing for other celebrations.

The two men I have so frequently mentioned during this description of my third trip to Cypress Hills, for some cause not known to me, appeared very anxious to be away from this Indian camp, and, as I shared in this anxiety, we were in our saddles by six o'clock and on the last stage of our journey. While the police officer and I rode ahead, our team and men followed. The trail was good and, although my Yankee attendant made a bold attempt to run me down with his horses and waggon—which almost cost him a bullet in his unworthy body—we reached Maple Creek safely, where, nothing loath, I gave up my outfit and prepared to reach the last locality I had been instructed to examine before returning to Ottawa.

Although I had been unfortunate in this, my last journey to the Cypress Hills, we secured several specimens new to our Canadian Miocene rocks, among which may be mentioned a left ramus of mandibule of an extinct wild boar. I think I have already mentioned that Professor Cope named this fossil jaw *Elotherium arctatum*. It is described and figured in "Contributions to Canadian Paleontology—Vol. III. It is probable

AMONG THE ROCKS.

further researches in the Miocene rocks of the Cypress Hills may reveal many other remains of extinct animals, other than those now exhibited in the cases of the Dominion Geological Museum.

Taking the C. P. Ry., Aug. 6th, I arrived two days later at Port Arthur, and at once commenced my researches for fossils in that portion of the Lower Cambrian rocks known as the Animikie series, which has an area of over 1,000 miles, extending from Port Arthur along the west shore of Thunder Bay, Lake Superior, and other localities in that region.

The rocks of the Animikie formation consist chiefly of even bedded grey and blackish argillites, diabasis and other traps, cherts and jaspers. In places, the argillites (clay slates) are very micaceous and stained with oxide of iron, some are soft and stain the fingers with a carbonaceous matter, others are hard and dolomitic. They are supposed to have a thickness of 12,000 feet and to rest upon the Archian formation. They form the silver-bearing rocks of Lake Superior. Among the group of mines now being worked in the Animikie formation are the well known Rabbit Mountain and Beaver mines, which are situated a few miles from Port Arthur.

Although the stratigraphical relations of the Animikie series seem to prove that they are not far removed from the Laurentian formation, they are thought by some writers to belong to a higher series of strata—a series in which one might expect to find fossils of a Primordial type, but, although the enormous exposures of these slates at the silver mines, Rossport, and other localities afford the collector such a splendid chance, no fossils have yet been found in the so-called Animikie series. For various reasons it is exceedingly important that fossils should be found in these rocks, and the geological or other investigator who does find organic remains, whether belonging to the animal or vegetable kingdom, will solve a most important geologic question.

Mountains, hills, and valleys—part of which are heavily timbered—characterize the Thunder Bay, Lake Superior silver mining district. Bold escarpments of trap and other eruptive rocks give a rugged aspect to portions of the country.

A three or four hours' stage ride carried me from Port Arthur to the noted Beaver and Rabbit Mountain silver mines. At the Beaver mines I found my old friend Capt. Williams, with whom I had spent many pleasant hours at the Echo Lake Copper mines. Capt. Williams, who was then in charge of the Beaver mines, invited me to accompany him through the workings. I at once embraced the opportunity, and after being enveloped in rubber boots, coat and cap, with a tallow candle in my hand, we were lowered into the bowels of the earth. Down, down we went till my head became dizzy, but I closed my eyes, and held on to the side of the cage till a series of sharp bumps under our feet notified us that we had reached the bottom. Just at that moment, far away in one of the tunnels, an explosion took place shaking the foundation on which we stood, but my guide said it was all right—his men were blasting.

One who is not nervous finds much to interest him in these silver mines. Here and there in the great walls of black slate were partly imbedded cannon-shot-like concretions : the miners call them "bombs." They vary in size from half an inch to two feet or so in diameter. In the dense darkness of this "silver pit" our tallow candles showed but a faint light, sufficient, however, to reveal the beauty of some of the trap and other rocks whose sides fairly glistened with iron pyrites, calc-spar and other minerals. But so little silver ore is seen that one wonders where the $5,000 worth of that metal, taken out last week, came from.

The vast accumulation of dump matter taken from these mines and deposited near the mouth of the pits affords the searcher for fossils an excellent opportunity, but in the many thousand tons of rock examined not the least trace of organic structure was seen.

Leaving the Beaver mines Aug. 14th, I again took the C. P. Ry. to Rossport, a small fishing station on the south shore of Lake Superior, about eighty miles east of Port Arthur. On my way I met Dr. Lawson, one of the field geologists of the Survey, who was then engaged tracing out the boundaries of the Animikie formation. Having secured lodgings at the only boarding house in that vicinity and hired a small fishing smack, together we visited several small islands within easy reach of the shore. On one of these, Quarry Island, sandstones of the Nipigon series are largely displayed and are seen to rest on the black slates of the Animikie formation. They resemble some of the Potsdam rocks of the St. Lawrence River, but, so far as we know, contain no fossils; they have been largely used in the construction of bridges, etc.

Another island is partly composed of Amygdaloidal traps, others display fine sections of the Animikie carbonaceous slates; one is not likely to soon forget a sail around these beautiful islands, nor the delicious supper of fish one can always get here—fish just fresh from the deep cold waters of an almost matchless lake.

Rossport—if one can do without the luxuries to be had in the smallest town—is a charming place to rusticate in for a short time during the hot summer weather. Boating, fishing, gunning, and a hundred and one shady nooks to explore, nooks which remind one of the song in the Opera "Wang,"

"A shady nook, a babbling brook."

AMONG THE ROCKS.

August 21st, I was again in Ottawa unpacking my gleanings, from a few of the rocky leaves containing the relics of that wonderful epoch—the Age of Reptiles. The great extent of Miocene Tertiary and adjacent Belly River series, displayed along and near the banks of the Red Deer River, have as yet only yielded up to us, so far, a few of the many palæontological relics they must contain. Let us hope that others interested in the past history of the great Dinosaurian age, will continue researches in these rocks—only a few drops in the bucket, as it were, of which have yet been partially examined, and yet have added so much more valuable knowledge to the story of our Canadian fossil vertebrates.

CHAPTER XIII.

THE "QUEBEC GROUP" CONTROVERSY—AN INTERRUPTION BY ILLNESS—I AFFORD AMUSEMENT TO THE NATIVES—THE MONTMORENCY ROCKS—THE GOLD-BEARING ROCKS OF NOVA SCOTIA—IMPROVEMENTS IN THE MUSEUM OF THE SURVEY—A CLERICAL GEOLOGIST—THE PILGRIMS TO BONNE SAINTE ANNE—THE GEOLOGY OF THE SAINT ANNE DISTRICT—SURROUNDED BY THE TIDE.

DURING the year 1889, the "Quebec Group" controversy, was still a bone of contention, and again I received instructions to examine—especially for fossils—certain rocks in various parts of the Eastern Townships beginning at St. Julie, a small village, a short stage ride from the Becancour Station on the G. T. Railway, from which place I followed the rock exposures to Inverness, Lloyd's Mills, through the village of Ste. Agathe and St. Sylvester to the Chaudière River, thence to Point Lévis (South Quebec.)

More than one person had said to me: "you geological fellows have a grand picnic all the summer, and spend the winter preparing for the next summer's sport"; well, if these grumblers had followed my footsteps even over this short journey they would have found there was not much sport in it, and "those geological fellows" well earn all the money they get for "picnicing."

AMONG THE ROCKS.

It would take too long to record all the interesting incidents of a geological trip through the Eastern Townships, so I shall only mention one which may be of use to those following up this work·

At Lloyd's mills, in a deep gorge, through which the waters of the River Thames rush over rugged escarpments of slate and limestone conglomerates, I found in the latter rocks, several examples of the pretty little fossil *Salterella rugosa*, (Billings,) a very small conical shell, belonging to the Sub-Kingdom Articulata. I found that these conglomerates had a dolomite matrix and resembled very much the dolomitic conglomerates of the west-end of the island of Orleans (over fifty miles away.) Now, I thought, when I reach Quebec, I will at once cross over to Orleans Island, and if I can find *Salterella* in the dolomitic conglomerates there, it will be an interesting link in the palæontological chain of the " Quebec Group." I reached Quebec Sept. 12th, crossed over to my conglomerates, and there—a few yards distant from where Sir William Logan and I had sat years ago—I found a beautiful little group of *Salterella*. I thought how Logan's eyes would have glistened had we found these fossils then! My little prize now occupies a place in the cases of the Survey museum.

A few days were spent on the rocks of Point

REMINISCENCES

Lévis and Quebec; and then I returned to Ottawa, having finished my field work for 1889.

The summary Report of the Geological Survey for 1889, after giving a brief account of my explorations for this year, says :—

"The remainder of the year has been occupied by Mr. Weston in preparing, labelling and arranging for exhibition in the Museum the collections received during the summer."

During my connection with the Geological Survey, to the year 1890, I had seldom been away from official duties, more than a day or two at a time, but this year was an unfortunate one for me. A severe attack of la grippe brought on another complication which taxed the medical skill of my friend, Dr. H. P. Wright, to whose constant care I probably owe my life.

The following item is from the Summary Report of the Geological Survey for 1890 :—

"From the 6th of January to the 26th of June, Mr. T. C. Weston was confined to his house by severe illness. He subsequently obtained leave of absence until the first of August, and during that time and up to the 8th of Sept. he was occupied in the examination of the rocks in the neighborhood of Quebec City, from which he was successful in obtaining a remarkable and in some respects a unique set of fossils, which throw considerable light on the age of these rocks, which is still under discussion. From the 10th of September until the close of the year, Mr. Weston's time has been spent in museum work in the palæontological and archæological sections, in arranging several hundred new specimens in the cases, in preparing descriptive labels for them and the like."

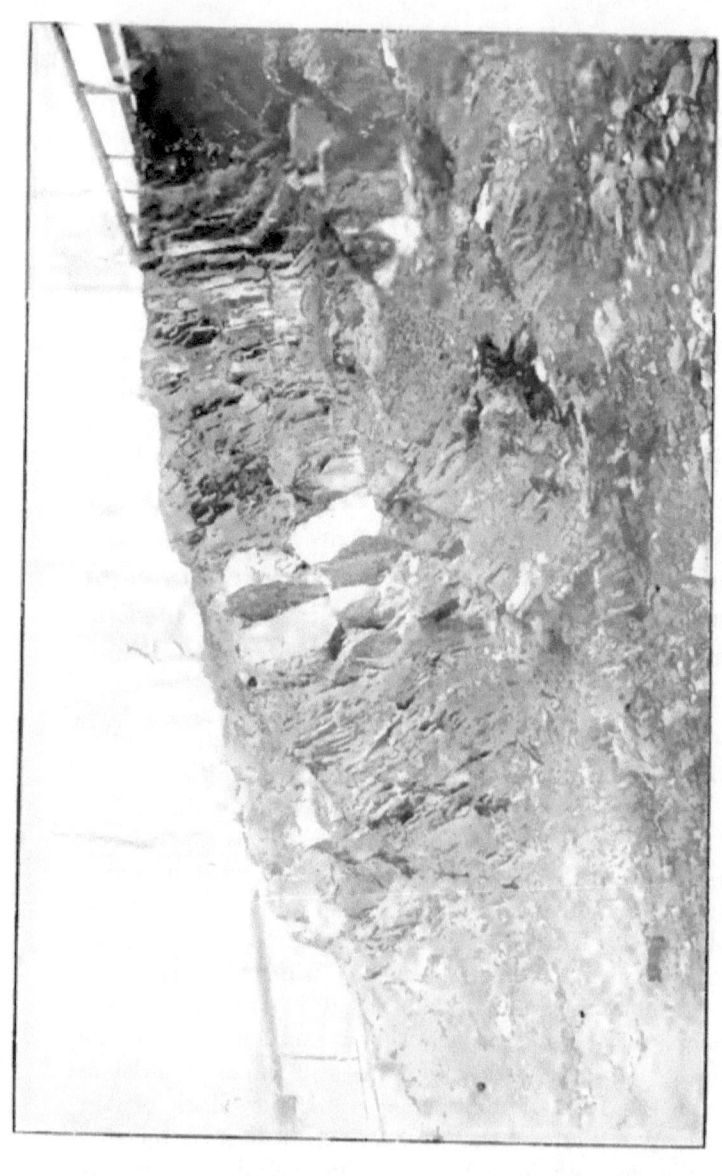

SHALES AND LIMESTONES, BACK OF ST. JOHN'S OR MONTCALM MARKET, QUEBEC CITY, 1800: Covered in 1897, while making local improvements. These are the rocks which yielded the remarkable fauna of fossils spoken of.

AMONG THE ROCKS.

The fossils from the Quebec City rocks, mentioned in the above quotation, were found in the grey slates near the back of St. John's or Montcalm Market.

It is a singular fact, that in all the geological work done by Logan, Billings, Richardson and others, up to this date (1890,)—and two or three years previous, when I discovered minute trilobites, small brachiopods belonging to the Lingulidæ family, and other forms in this same series of slates,—no fossils had been found, so far as I know, in the city rocks. These trilobite and Lingulæ beds have long been under the roadbed of the improved portion of St. Patrick's Street, and now while I write (1897) the rocks in which the remarkable series of graptolites, bivalve and other fossils were found in 1890, are being covered up in making local improvements; we regret that we did not do more work at these rocks—now "Lost to sight, but still to memory dear." Many of the fossils just referred to were obtained under rather trying circumstances. The "St. John Market rocks"—now covered—are in a thickly populated part of the city, where the youthful French element is almost unlimited. To these juvenile members of (the "Quebec Group" I was going to say,) the ancient city, I owe more than one blow which might have been serious. To see a man standing on a ledge of rocks, hour after hour in the blazing sun pounding these rocks, and every now and then carefully wrapping up and bagging the pieces, was —so said one boy in French—"as good as a circus." One day, a crowd of rough boys returning from school came down the bank in a body to interview me. I tried to explain what I was doing, but my little lecture on palæontology evidently did not please them, for they left in a body, ascended the bank, and opened fire on me with corn-cobs and other vegetable matter of which there was plenty at hand. I was getting the worst of it when fortunately an old priest came along, and seeing the trouble, came to my assistance and rescued me from the hands of the Philistines.

Collecting "diamonds" to sell to the many tourists who visit the "Ancient City" is a favorite pastime for many Quebec children, and as my "market rocks" yielded many specimens—which could be sold for ten cents or so a dozen—they were a perfect mine of wealth to the little ones who on half day holidays gathered here armed with any iron implements, from a railway spike to an old horse shoe, indeed one little girl confessed that she had stolen her father's shoemaking hammer for this work. Seeing me, one day, working in a certain spot for a long time, a number of these children, concluding I had made a good strike, crowded me so much that I was about to give up work when the thought

AMONG THE ROCKS.

occurred: Suppose I offer these youngsters five cents a dozen for all the diamonds they can get anywhere ten yards from where I am working. The offer was made and gladly accepted, and for two hours I worked in peace; then at five o'clock I called time, and the youngsters gathered around me with their "diamonds," some had two or three, while others had many. I found it no easy task to settle up accounts with these young folks, but at last matters were squared satisfactorily, and I went away with about five dozen "diamonds," which after all cost me fifty cents. Now, when I look at the fossils collected that day, I seem to see those twenty or thirty children pounding away at the rocks—some singing French songs, while another small group sang—

"Digging dusky diamonds all the season round,
"Down in a coal mine underneath the ground."

Passing two young women in Quebec the other day, one who carried a baby nudged the other and said: "That's the fellow that bought our diamonds, when we were youngsters, wasn't them happy days?"

This year (1890), owing to differences of opinion regarding the rocks of Montmorency Falls, it was decided to have each variety of these rocks examined microscopically. This duty was assigned to the lithologist of the Survey, Mr. Ferrier, and I was requested by the director to make a collection of all the typical rocks of the

REMINISCENCES

falls. I was still very shaky, after my late illness, and this task, although accomplished in one day, was a severe one. However, a very interesting stratigraphical collection was made, which I regret to say has not yet been reported on.

While trimming "hand specimens," a short distance up the river from the falls, a lady came along with her "kodak," and after taking several snap-shots at the beautiful scenery of this delightful spot walked away in the direction of the hotel. In a short time I too made my way to the hotel, where I learned that this lady was the Countess of Aberdeen, who with her noble husband the Earl of Aberdeen was, I believe, making a flying visit to Quebec. A few years later (1895) Lord Aberdeen became Governor-General of Canada. To their Excellencies the people of Ottawa will always be indebted for their good works, delightful winter sports, and other social gatherings at Rideau Hall.

During the winter of this year (1890), the geological horizon of the gold-bearing rocks of Nova Scotia afforded considerable discussion. Certain nodular forms, which when weathered seemed to show coralline structure, had been found by Mr. Willis in the rocks of the Northrup gold mines of Nova Scotia. These forms were handed to Professor H. Y. Hind, who supposed them to be fossils of Lower Silurian age. They were then handed to Professor G. T. Kennedy, of

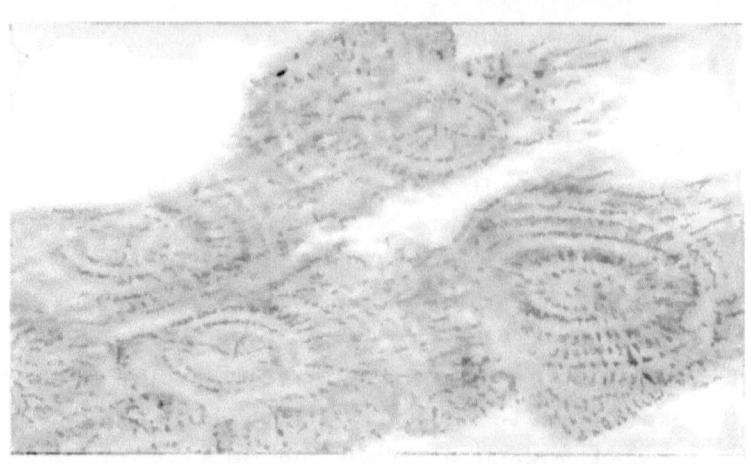

HURONIAN CONCRETIONS: Once supposed to be the fossil *Oldhamia radiata*. From the Huronian argillites, Newfoundland.

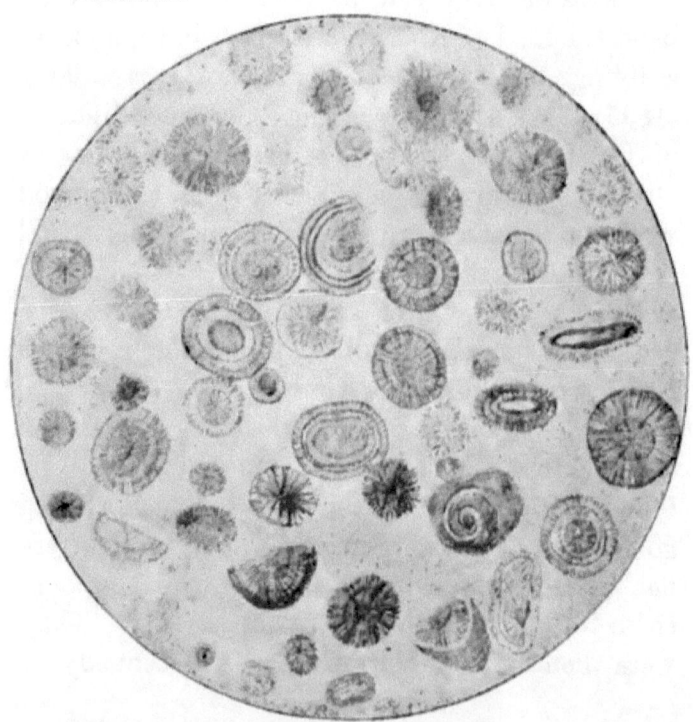

CONCRETIONS: Micro-drawing from a thin slice of Cape Breton oolitic limestone, magnified about twenty times.

King's College, who confirmed Professor Hind's opinion, and pronounced these "fossils" to be *Stromatopora*. Had this been a fact, it would have altered our present idea of the stratigraphy of the Nova Scotia gold-bearing rocks, considerably. Mr. Fletcher and Mr. Faribault of the Geological Survey having collected numerous examples of these supposed fossils, I devoted considerable time preparing microscopical sections, etc. of these forms; the result of which was that not a particle of organic structure was found, and like the wonderful discovery of *Oldhamia* in the Huronian slates of Newfoundland—mentioned elsewhere—were found to be only concretions.*

A considerable time was spent this winter in hunting up the dates when certain old museum specimens were collected. Although the localities were on these specimens, for some reason the date of collection was omitted. It is due to Dr. Selwyn, then director, that now almost every specimen in the museum bears the date of collection. This and other museum work—together with a large share of time spent in the restoration of fossil bones, the developing of other specimens, etc. kept me fully occupied till the spring. Before closing my remarks for this year I take much pleasure in recording my sincere thanks to

(* See notes on concretionary structure—by T. C. Weston—Trans. of the Nova Scotian Inst. of Science—Ser. 2, Vol. 1.)

REMINISCENCES

Dr. H. M. Ami, M. A., F. G. S., of the Survey, who during my sickness—besides his own work—devoted much time attending to many of my duties, both in the palæontological and ethnological branches of the Survey, and later in assisting me to select typical graptolites of the Point Lévis and Quebec City rocks, to send to Prof. Walcott, director of the Geological Survey of the United States, and Prof. Lapworth, LLD., F.G.S. of Birmingham, England, who had kindly consented to identify these fossils. Prof. Lapworth's interesting report on these Quebec graptolites will be found in Trans. Royal Society of Canada, Section IV. 1886.

As will be seen by the reports of the Geological Survey, the year 1891 was an interesting epoch in the history of Canadian geology and also in the history of our National Museum.

It was my privilege to be allowed to make many improvements in the arrangement of specimens, both in the palæontological and ethnological branches, to add many new specimens to our already splendid geological exhibit, and to restore many of the fine fossil bones which now form one of the interesting features of the museum. In this work the winter months passed pleasantly. The "Quebec Group" controversy continued, and more than one unbeliever in our director's stratigraphy of the Lévis formation got an occasional roasting!

AMONG THE ROCKS.

To try and prove certain theories by fossils I was again preparing to leave for further work at the Quebec City, and island of Orleans rocks, when Ottawa—in fact the whole of Canada—was cast into mourning, by the death of our Premier and Minister of Railways The Right Hon. Sir John A. Macdonald, which took place on the 6th of June. By the death of this great man Canada lost its greatest statesman, and the Geological Survey one of its best friends.

On the 30th of June, I was again in the City of Quebec, spending a few days on the St. John's St. " Market rocks"—which now, as I have already stated are covered up. It was during my work this summer, that a very interesting series of fossil *Graptolitidæ*, belonging to the genera *Dictyonema*, were found in one of the layers of slate composing part of these " Market rocks." Now that these rocks are for ever hidden from us, the fossils obtained from them with so much trouble and perseverance should be considered among the choicest specimens of the geological Survey.

During my visit to Quebec, I have often spent interesting half hours with Abbe Laflamme of Laval University, who, as the reports of the Survey show, has, during his vacation from the University, done much important geological work for the Survey. Like all true geologists, the professor had a sharp eye for fossils, but in some way, he like others seems to have taken it for granted that the Quebec City rocks contained no fossils, an idea which might be due to the fact that thousands of tons of the black limestone of the city may be broken without finding the trace of a fossil. Having on the 6th of July, mentioned to Prof. Laflamme the discovery of *Dictyonema* and other graptolites, in the "market rocks," he at once locked his study in the University, and accompanied me to these rocks whose fossil evidence had so long escaped the notice of geological workers in the "Quebec group" formation.

To see a black robed priest, with hammer in hand, pounding rocks in the heart of the city must have seemed strange to many passers-by, but even the curiosity of the small boy was suspended during the priest's stay with me. Had I been alone no doubt there would have been a few corn-cobs flying around as on former occasions.

During my few days sojourn in Quebec this summer, I had the pleasure of taking my friend Mr. R. G. McConnell, of the Canadian Geological Survey—who, during a vacation, was making a hasty trip to England—over a portion of the Quebec City and Lévis rocks. McConnell, who is one of the "long distance" Northwest explorers of the Survey, had never been in Quebec before, or seen in *situ* any of the Quebec or Lévis rocks, so often discussed (sometimes hotly) in his presence in the work-rooms of the Survey. To a

AMONG THE ROCKS.

veteran geologist a different series of rocks from those he has been studying for years always proves interesting; so it was with McConnell's hasty glance over a portion of our Quebec group rocks.

Accompanied by Mr. Bedard—a Laval University student—as assistant, we boarded one of the market boats which runs from Quebec to several small villages along the north side of the island of Orleans. In a few hours we landed at St. Famille—the small French village spoken of while recording a previous journey with Dr. Selwyn some years ago. From this village we went carefully over the shore rocks, both east and west, hoping to gather new facts regarding the relation of these rocks to those of Quebec city. After much hard work and weary tramps nothing was found to disagree with Logan and Richardson's interpretations of these rocks. The only new feature recognized this summer was that near the west point of the island of Orleans, nearly opposite Beauport church, there is a set of black limestones containing fossils identical with some of those found in the St. John's market rocks. This proves, without doubt, that these Orleans Island limestones are a portion of the black rocks seen in most parts of Quebec City.

To try and confirm, by fossil evidence, some of the statements made regarding the geological horizon of the rocks on the north shore—opposite the island of Orleans—I proceeded alone by boat, to Ste. Anne de Beaupré—a village situated on the banks of the St. Lawrence River—21 miles from Quebec. But who has not heard of that wonderful spot where thousands—yes, hundreds of thousands—of pilgrims assemble yearly to do homage at the shrine of St. Anne, to hear mass in the magnificent church, and drink water at the " miraculous " fountain. Rich and poor, old and young, kneel side by side at the beautiful altar, and are blessed by one of the numerous priests who are ever ready to administer spiritual consolation to the afflicted, many of whom make heroic sacrifices to reach this wonderful place of miracles :—

> " The waters of the grand Saint Lawrence glide
>   In calm majestic motion, on their way
> Past *Bonne Saint Anne*, as if the gentle tide,
>   Its silent humble homage thus would pay
> Before the ancient shrine, as on its breast
>   It bears the pilgrims to this place of rest.

At this time, 1891, there were no large hotels, as now, at Ste. Anne, and I found shelter in one of the many cheap boarding houses of this strange place of religious associations.

A great many people who visit *Bonne Sainte Anne* for religious comfort are poor indeed! a fact readily seen when one looks at the hundreds of rudely made crutches, trusses, and other instruments, constructed to support the crippled and otherwise afflicted. Year by year these instruments of suffering humanity have accumulated. Day by day these relics of pain and decay are laid at the shrine of Saint Anne. Tobacco and snuff boxes, pipes, spectacles, jewellery, and many other things have been laid with grateful hearts, for health restored and hopes renewed, at the feet of Saint Anne. Hundreds of these offerings adorn the walls of the church, while other large instruments made for the afflicted cover one of the great pillars of this noble building.

But to return to our boarding house, where, to reach the small bed-room assigned me, one has to pass through a large sitting room which, at this time, was used by "women only," pilgrims who come and go every hour of the day. Having wended my way past twenty or more chattering French Canadians and gained my " chamber," I prepared at once for a good geological tramp, then sought the dining room where, seated on long benches before a large, roughly made deal table, were a motley group of pilgrims. Each of these visitors to Ste. Anne appeared to be eating provisions they had brought—some brown bread and a little pork, while others seemed to fare luxuriously on sweet corn, bread, butter, and cakes, purchased in the village. Some of the more fortunate ones indulged in a plate of soup, or a cup of tea, furnished by the house for five cents each. But poor and scanty as the food of many of these people seemed, they were all happy, and why not? They had made their oblations to the good Saint Anne, drunk of the water of the "miraculous" fountain, been blessed by the good priest, and were now waiting for the steamboat to carry them back to their homes.

After dinner—which was taken at the same table with the pilgrims, I hurried off to my work, returning late in the evening to find the house crowded with a new lot of pilgrims, who for some reason had been detained on their journey and were now obliged to remain over at Ste. Anne for the night, an unusual thing for pilgrims to this place.

The night was beautiful, and after supper I wended my way down to the foot of the long wharf, over which, during the last year (1890), 105,672 pilgrims had walked on their road to do penance and receive Holy Communion at the shrine of Saint Anne, to do which many had made brave sacrifices.

Alone I sat in the gloaming on the foot-worn steps of the pier and watched the thousand twinkling lights of the Quebec and Point Levis shores, and their reflection in the dark waters of the St. Lawrence.

From this distant point one sees a great portion of the district surroundings of the ancient capital of Quebec, every yard of which is freighted with historic interest, the cliffs of the Citadel, the magnificent Dufferin Terrace—1,400 feet long and 200 feet above the level of the river, the towering cliffs of Lévis which have yielded us such an abundant fossil fauna, and, after researches made almost every summer for the past fifty years, still give us an occasional new form, but—

> "The day is done, and the darkness
> Falls from the wing of night,
> As a feather is wafted downwards
> From an eagle in his flight."

And again I wended my way to the pilgrims' lodging house, passing the wharf toll-gate through which 2,000 seekers after religious consolation had passed that day, each paying a small toll, which, to save time, is arranged for by the good fathers who organize the excursion.

A short walk and I was again in my boarding-house. All was quiet, and unobserved I reached the large room that led to my bed-room. On opening the door, to my astonishment I saw the floor covered with cheap mattresses or straw beds on each of which was stretched, one, two or three women. I backed out of the room, closed the door and went in search of the good woman of the house, and having found and explained matters to her, she smiled and said—"Monsieur, I will escort you to your chamber." Following my hostess, we picked our way carefully between these sleeping pilgrims and gained my chamber.

REMINISCENCES

I was tired and slept well till late the following morning, when on opening my door I found all quiet again, and the people of the house busy preparing for another batch of visitors to this village of miracles. And so these pilgrims come and go every day during the summer months, each leaving their small oblations towards the support of the church, and each carrying away some small souvenir of their visit to the shrine of Saint Anne.

It is a beautiful sight, while leaning on the rails of Dufferin Terrace at sunset, to watch one of our large St. Lawrence steamers heavily laden with homeward bound pilgrims glide gracefully past the walls of the Citadel and through the shadows of the towering cliffs of Cape Diamond; to watch the brilliant reflections of the evening sun on the waters, the cliffs and the numerous windows in the Catholic buildings of Levis. But hark! a thousand voices on the throbbing steamer are chanting their vesper hymns. Happy souls, many of whom endured much to make this pilgrimage to their *Bonne Sainte Anne*, to

"Confess their sins, receive the Bread of Life."

But to our geology again. Returning to Ste. Anne by the Beauport and Montmorenci road we pass numerous outcrops of Trenton limestone and many small lime kilns in which the French Canadian farmers burn lime for agricultural purposes and for whiting the walls of the houses.

AMONG THE ROCKS.

At the summit of Montmorenci Falls—of which I have already spoken—the Trenton limestone is beautifully displayed in the "natural steps," of the Montmorenci River. Here the formation has an estimated thickness of fifty feet of dark grey limestone (resting on gneiss) holding most of the typical fossils, among which are found good examples of our Lower Silurian Bryozoa, Brachiopoda, Gasteropoda, Cephalopoda, Crustacea and other divisions of the fossil fauna of the Trenton formation.

An interesting geological feature of this delightful locality—in the shadows of pine woods, and the roar of waters as they hasten on to hurl themselves over the great cliffs and mingle with the waters of the St. Lawrence—is the occurrence of a rich deposit of ochre, covering an area of several acres, and having a depth of one to seventeen feet. The colors of these pigments vary from a light yellow to a dark brown. In 1875 while making a collection of these ochres for the Philadelphia Centennial Exhibition I found a number of flint chippings and bits of Indian pottery. Some of these relics were a foot or more below the surface. It is probable this was a favorite Indian camping ground during their hunting expeditions—hundreds of years before the first page of Canadian history was written.

Continuing our journey by road, from the Falls of Montmorenci; towards the Ste. Anne

River and at the lower falls, almost all the typical rocks of the Trenton, Utica and Hudson River formations. These have characteristic fossils and a great variety of strata of bluish-grey limestone, dark bituminous shales, grey and other colored sandstones, etc. Together with the Laurentian gneiss on which one of these formations (Trenton) is seen to lean or rest, they form a thickness of about 1200 feet.

Arriving again at Ste. Anne I take another meal at our pilgrim boarding house, then proceed by buggy to St. Joachim another small French village on the banks of the St. Lawrence River, a few miles below Ste. Anne.

It is a lovely drive, the river to the right and verdant sloping banks or terraces to the left. Outcrops of sandstone and other rocks belonging to the Hudson River formation are numerous. Most of these I examined carefully for fossils. In one ledge of sandstone good examples of the Graptolite *Diplograptus pristis* were found. These were important specimens, and I was intent on my find when a typical lower-class French Canadian of this district who was passing stopped to see what I was doing. Seeing me carefully wrap up and bag a piece of this fossiliferous rock, he turned to my carter and said :—

"Sacré crapaud! pourquoi emportes—tu dans ton sac, ce petit morceau de pierre ? Mon Dieu!

AMONG THE ROCKS.

il peut bien prendre, toutes les pierres qu'il trouvera sur ma ferme!

Continuing our journey in a short time we came to the Grand River, a small but rapid stream—hurrying on to plunge and mingle its waters with those of the St. Lawrence.

Here where the road crosses the stream, we find a beautiful example of rock crumpling or folding, forming a number of small anticlinals and synclinals—But to which formation do these strata belong? They look like some of our Lévis rocks; and they ought to belong to the Hudson River formation—what will the Index to our stony book say? With hammer in hand I wade along the margin of the stream to where it cuts through a series of black shales and thin bedded sandstones. A few minutes work and I have in my bag several typical fossils of the Hudson River formation. The stratigraphy of our rocks across the river—on the north shore of the island of Orleans—told what we might expect here, and the fossils in my bag confirm the supposition.

Quite satisfied, and the best of friends, with this fine example of corrugated strata and the fossils obtained from some of the beds, my carter and I sat down in a shady nook on the river bank to enjoy a lunch the good lady of our pilgrim boarding house had provided us with—a bottle of milk four hard boiled eggs, bread, butter and cheese,

REMINISCENCES

we were both hungry and it was a delicious meal—so my carter thought—and when I gave him a cigar, he murmured words of thanks in which *le Bon Dieu* figured! Man and beast being fed and rested, we resumed our journey, and when within a mile of St. Joachim Church came to an interesting outcrop of dark grey limestone mentioned in the Canadian Geological Report for 1863. Here we do not have to hammer for hours—as in some cases—for a fossil, for the index of this chapter of our stony book is opened out before us, and we see at a glance that the fossils exposed on the weathered surface of this sedimentary rock are characteristic forms of the Trenton formation.

A few minutes suffice to give us all the typical forms we need to identify this strata, some of which belong to the following genera: *Stenopora, Leptæna, Strophomena, Orthis, Calymene,* etc.

Between this Lower Silurian strata and St. Joachim village only Laurentian rocks are seen, but the limestone is met with again at Bay St. Paul, fifty miles or so lower down—a locality I shall mention in my next year's (1892) rambles

St. Joachim village, like St. Anne, and I think all the villages along the lower St. Lawrence coast, are essentially French Canadian settlements. In every village a large Catholic church occupies a prominent position. They are all dedicated to some saint. " In the gospel according to St. Luke, Joachim is mentioned under the abbreviated name of Eli, as father-in-law to St. Joseph. The only but glorious offspring of this marriage was the Blessed Virgin Mary." However large these churches are, they are seldom large enough to accommodate the great number of people who assemble for Sunday morning mass. Sunday is a red letter day for the toiling Canadian farmer and his family. To some who have to make long distances to their church, it means most of the day at mass and on the road ; but after their devotions are over, the remainder of the day is given up to pleasant intercourse with one another, and for a short time the trials and cares of life are forgotten. After seeing much of the poor class of French Canadians one comes to the conclusion that they are indeed a happy race of people.

Leaving the village of St. Joachim with its grassy flats where the "sport" finds good work for his dogs to do, we return to Ste. Anne, where again a few more geological tramps were made along the shore and shore road. Although the rocks had yielded many interesting fossils, I found nothing to refute the work done by Logan and his colleagues, during the early history of the Survey.

July 20th, my friend Mr. Bédard joined me in Quebec and together we spent a day on the

REMINISCENCES

Utica shales between the mouth of the St. Charles River and Beauport.

The Utica formation, which takes its name from Utica, New York, is represented in Canada by about 300 feet of black bituminous shales, interbedded with harder rocks, sandstones, limestones, etc.

At low tide near the mouth of the St. Charles River a fine section of this strata is seen, its dip is about S.E $<40°$. To the palæontologist who is interested in Utica and Hudson River fossils—the latter formation in Canada has a thickness of 2,000 feet and is intimately associated with the Utica deposits—this locality presents a fine field for collecting most of the typical Utica and Hudson River graptolites. It is singular though that while much of the strata here is very prolific in graptolites, I failed to find one of the trilobite or fossil bivalve shells, which are so numerous in the Utica of other localities.

The tide rises rapidly over the Utica rocks and extensive marshy flats near the mouth of the St. Charles River. This I found to my cost one day, for while intently examining some fine specimens of the graptolite *Climacograptus bicornis*, I felt a splash of water over my boots, and looking up, I found myself surrounded with water; the twenty or more cows who were near me a short time ago feeding on the marsh grass, had evidently paid more attention to the tide

AMONG THE ROCKS.

than I, for they were on the shore half a mile away. Slinging my collecting bag, which contained my lunch and perhaps fifty pounds weight of rocks, over my shoulder I made for the shore, which I reached just after the water had found its way into my trousers pockets. The day was very hot, and while my clothes dried in the sun I watched the burning of a schooner which had been beached here for that purpose, and thought of those startling military times of 1759, when the siege of Quebec took place and all this shore between the mouth of the St. Charles River and Beauport was lined with batteries, and the surrounding country for miles was covered with French encampments.

A spot a little to the right of where I sat enjoying a bottle of ginger ale and some sandwiches, provided me by my good sister-in-law, Mrs. M. G. Mountain,—an old resident of Quebec—is still pointed out as the place where in 1535 Jacques Cartier and his companions passed the winter. It was here also that the French constructed floating batteries and fire-ships which floated with the current down the St. Lawrence and did much damage to the English fleet.

July 22nd, I was again on the island of Orleans trying to connect by fossil evidence the black shales and limestones of the "cement quarry" with the bituminous limestones and slates of Quebec City. The fossils obtained

REMINISCENCES

threw much light on this complicated strata, but much still remains to be done. I returned to Ottawa July 27th, and owing to delicate health did not return to field work again this year.

During the remaining part of the summer and winter of 1891-92 my time was occupied again in palæontological and archæological work connected with the museum, varied with microscopical and official work.

During the past winter my health had been poor and my request to be allowed to make short geological excursions during the summer of 1892, was readily granted.

Knowing that extensive excavations for the foundation of the Chateau Frontenac were in operation, and hoping to obtain more fossil evidence regarding the Quebec City rocks, I again on the 18th of July returned to the ancient capital.

CHAPTER XIV.

FURTHER GEOLOGISING AT QUEBEC CITY—EXAMINING THE ROCKS PRECIPITATED BY THE LAND-SLIDE—A VICTIM OF POISON-IVY—A SEARCH FOR A METEORITE ON THE OTTAWA RIVER—THE LAURENTIAN MARBLE—AN ARTIST'S SKETCH IS USEFUL—THE JOGGINS COAST, N.S. AGAIN—CONCLUSION AND THANKS TO MY FRIENDS OF THE SURVEY.

THE thousands of tons of black limestone excavated for the new Frontenac hotel, Quebec, afforded an excellent opportunity for fossil researches, and I worked hard to get new facts, but regret to say that with the exception of one species of graptolite—*Dyplograptus pristiniformis* these rocks were destitute of organic remains.

Returning to the St. John St. market rocks I found my *Dictyonema* beds gone; gone to pave the streets of the city. I have already stated that the road-bed of St. Patrick's street covers my trilobite beds, now the *Dictyonema* beds only a few yards distant, are gone. It is a singular fact that the fossil *Dictyonema* has not so far been found in any other than the St. John Street market rocks, in all the city. The loss of these rocks—which have been made historical by the writings of Professor Lapworth of Birmingham,

England, and several members of the Geological staff—will always be regretted.

While still at work at these ever interesting Quebec and Lévis rocks I received official instructions to proceed to Bay St. Paul, there to make a large collection of the garnetiferous gneiss which forms one of the interesting geological features of the north-east side of the bay.

Bay St. Paul which is one of the lesser salt water resorts of the river St. Lawrence, is about 60 miles below Quebec, 30 miles above the popular summer resort, Murray Bay, and 90 miles below Quebec. Between the latter Bay and Bay St. Paul, the coast is very rugged, and affords an excellent opportunity for the student in geology to study Canadian Laurentian rocks, and also some of the Lower Silurian strata; as at Bay St. Paul, where the Trenton limestones are largely developed and contain the characteristic fossils of that formation.

A little saline spring issues from the Silurian rocks of this bay, and a short distance away, in the parish of St. Urban—large deposits of iron ore (limonite) occur.

The journey from Quebec to Bay St. Paul in one of the fine Saguenay steamers occupies about five hours. Before reaching Riviere Ouelle, about 80 miles below Quebec we enter salt water and see, perhaps, several white porpoises, that peculiar Cetacea, white whale or

Beluga Delphinapterus catodon, which as a skeleton in the Geological museum, Ottawa shows a length of 18 or 20 feet, and fossil bones, also seen in the same institution, from the Post Tertiary clays of Montreal, show that this aquatic mammal existed prior to the formation of the Leda clay deposits in which such a variety of "chalk shells" are found.

A Quebec paper states that during this spring (1898) large droves of porpoises have been seen in the St. Lawrence following the small fish on which they feed. Nearly one hundred have been captured at the River Ouelle fishery. This catch represents a value of over three thousand dollars. One of these porpoise, or white whales, was sold to an American, representing the Barnum Museum for $80.

August 9th, I was again in Quebec accompanying Dr. Selwyn, director of the Geological Survey of Canada, over some of the Quebec City rocks he had not yet studied. We first examined the great pile of rock debris which fell from almost under the foundation of part of the Citadel walls in 1889, crushing through several houses and hurling fifty-one souls into eternity.

The falling of these rocks brought to light several beds of strata exhibited in no other part of the city, thus affording an excellent opportunity for fossil researches; but although we climbed over thousands of tons weight of this land-slide we failed to find more than one species of fossil, that being a graptolite common to the Hudson River and Utica formations.

From these fallen rocks we ascended the steep partially verdant cliffs till we were immediately under the walls of the Citadel. It was an exceedingly hot day, and the journey up about 180 feet of slippery shingle was not easy. When about half way up we came to a large bed of poison ivy; turning to me the doctor said, "Weston is this poison ivy?" "Yes" I said, "put on your other glove and if possible don't touch it. For my part I have no gloves nor umbrella, and must hold on to this villainous plant or perhaps roll to the bottom of the cliff." I had suffered from this poisonous plant before, and knowing that I was then in a fit state to receive the poison felt sure I should suffer again, but there was no help for it, so I used the stems of the plant as a ladder. Reaching the top of the cliff, without thinking where my hands had been, I brushed the perspiration from my forehead. The following day I took my chief to see the Utica slates, near the mouth of the St. Charles River (already spoken of). Before leaving Dr. Selwyn that evening, I held out my wrists and said, "look doctor, the poison is beginning to do its work." The next and following seven days I was in bed and suffered—well I shudder now when I think of the awful distress

AMONG THE ROCKS.

that infernal plant can cause—large water-blisters over one's face, one's hands tied up in linen saturated with lotions, and all the rest of it. It is a singular fact, however, that this poisonous plant, which covers a large portion of the Citadel rocks, only affects some people, and that only at certain times of the year, and when the skin is in a fit state to receive the poison. We are told in classical mythology that ivy was sacred to Bacchus, whose head was surrounded by a wreath of it. I wonder what the effect would have been had they crowned him with the same species of ivy that caused me so much trouble! The species of ivy spoken of above is *Rhus toxicodendron*. French Canadians call it—Herbe à la puce. It is found in abundance under the rampart and Citadel wall, along the shores and on the islands of the St. Lawrence river. Although in handling this plant it may only affect one person in fifty, it is better to avoid touching it in the hot months of summer.

During my stay in Quebec this summer (1892) I found an opportunity of again examining the Mountain Hill Cliff. Geologically this is a very interesting portion of the Quebec City rocks; but here too the moment one touches the rocks with his hammer, he is surrounded by people who want to know why you break the rocks while there is so much already broken, etc., etc.

REMINISCENCES

I was delighted one day when Mr. G. A. Lafrance, whose book-binding establishment comes close up to the cliff, allowed me to get out of a back door on to the escarpment. I worked without interruption, and found such important information regarding the Quebec city rock deposits, that some days after when the director of the Survey returned to the city, he gladly accompanied me, and together we crept through the little back-door, or trap, in Mr. Lafrance's factory. We left this locality with quite a number of fossils which prove that a large part of these Mountain Hill rocks belong to the Trenton and Hudson River formations.

The following extract, from a paper contributed by me to the Ottawa Naturalist, will serve to record in these memoirs, several interesting facts regarding this portion of our Quebec City rocks:—

" Out of the 12,000 feet or more of strata which form the much discussed " Quebec Group," there are several interesting escarpments and sections which have hitherto not received the attention they deserve. One of these escarpments is the Mountain Hill cliff, which forms a portion of the height over which the ramparts of the city of Quebec are built.

" A close examination of the cliff immediately facing Mountain Hill House, on the lower part of the hill, shows it to be composed of a coarse grey nodular limestone ; in places, bedded structure may be seen while the principal portion (which is the matrix of the conglomerate), is compact, and sometimes flinty, with seams of carbonaceous or bituminous matter.

AMONG THE ROCKS.

"This portion of the cliff is prolific in fossils, but they are chiefly fragmentary, and might readily be overlooked. This is probably the reason why, in the early study of the geologic structure of the city portion of the 'Quebec Group,' these were included in the Levis division of the same.

"No fossil remains had been found or observed in the Mountain Hill cliff until the summer of 1877, at which time the writer discovered a number of interesting species. In 1892, another opportunity was afforded me to examine that portion of the exposure immediately back of the Express office, and adjoining the book-binding establishment. On that occasion there were found some remarkably well-preserved fossils, some of which were immediately recognized as being characteristic Trenton forms. Dr. Selwyn arrived in the city at the time, accompanied me to this locality, and several new species were added to our former collection."

My last official journey of this season (1892) was in search of a meteoric stone, or large piece of meteoric iron—which celestial bodies, we know, occasionally fall from the heavens.

On the 3rd of September, Mr. Robert Abbott, of Ottawa, who has spent many years on the Ottawa River, mentioned to the director of the Geological Survey that some years ago he had seen a "chunk of iron," which he supposed to be a meteorite, on the banks of the Ottawa River near the village of Portage du Fort.

That same evening Mr. Abbott and I left Ottawa in search of this specimen. After five hours' railway journey we arrived at Portage du Fort, where, on seeing the rock deposits of that locality, I said "if anyone wishes to build a marble palace on the banks of this beautiful river, there is plenty of material here."

Although the Laurentian crystalline limestone of Portage du Fort, and other localities on the Ottawa, is not fit for statuary, it is well fitted for any ornamental work, and has been largely used for decoration purposes in the Parliament buildings of Ottawa. The following morning (Sunday), Mr. Abbott and I started after the "thunder-bolt," arriving at the spot where he had seen the "chunk of iron" some years ago. We found we had—as Mr. Abbott expressed it—had a wild goose chase. We searched diligently among the Laurentian boulders which cover a large portion of the bank of the river here, but failed to find the "Thunder-bolt." Returning to the village, we made enquiries from some of the old habitants of the place. One said that some boys, having heard that the "iron stone" was valuable, rolled it into the river till such time as they could find a way of disposing of it. Another said that priests from Montreal had attempted to cart it away, but finding they could not do so with the means they then had, buried it till they could find a way to transport it to the railway. Mr. Abbott affirms the specimen was there, and that he had seen it many times, each time thinking to report the matter to some public institution. It is a great pity he did not while it was there. However, he still hopes to find this valuable specimen.

The large and only meteorite in the Ottawa geological museum now, is said to be of extraterrestrial origin and an aerolite. It was found in 1854 upon the surface of a field at Madoc, and weighs 370 pounds—so says the Geological Survey Report for 1863. It is probable that this specimen was first found on the surface of a field; but Sir William Logan told me that he found it propping up the corner of a barn, and at once sought the owner of the barn and offered to put a good square stone in its place; the offer was accepted and Sir William immediately had this valuable specimen removed and placed in the museum.

My official duties during the winter months of 1892 consisted as usual, in museum work connected with the palæontological and ethnological branches of the Survey. During the first three months of 1893, some of the members of the geological staff were busily engaged preparing maps, rocks—fossils, etc., for the forthcoming World's Fair at Chicago I was requested, as usual, to prepare a series of specimens, illustrating, as far as possible, the supposed Laurentian fossil *Eozoon Canadense*—already spoken of in these pages. The preparation of these specimens occupied about two months. They consisted of decalcified and other specimens showing the tubuli structure,

REMINISCENCES

microscopic sections, micro-photographs, micro-drawings, etc., altogether forming one of the most interesting collections of *Eozoon* yet exhibited.

My next duty was to prepare a stratigraphical collection of Canadian fossiliferous rocks. These were included in the large collection of hand specimens which represented as far as possible examples of all our Canadian rocks.

My next work, also for the coming exhibition, was the re-labelling and re-arranging a large number of specimens of native gold, models of gold from the Chaudière River, specimens of platinum, etc., and lastly in preparing a series of labels, tablets, etc., to illustrate the methods employed by the Survey in labelling and exhibiting geological specimens in the museum. These specimens filled three of our museum flat show cases, which when in place in the Chicago Exhibition formed an interesting and instructive feature of our geological exhibit.

June 17th, 1893, through indisposition I obtained a short leave of absence and again returned to Quebec. After a few days' rest, although the flesh was weak, the rocks on the other side of the river had such magnetic attraction, that again, with a boy to hammer, and carry my collecting bag, another important collection of fossils was made.

AMONG THE ROCKS.

I think I have already mentioned that fossils are not evenly distributed through the Lévis rocks, but occur generally in colonies; so one making his fossil debut to our Quebec city and Lévis rocks must not be like the foreigner who said to me: "I have tapped the Levis cliffs, and find them to be Laurentian, for there is not even the ghost of a fossil in them."

Expecting to visit Murray Bay for a few days, I had, before leaving Ottawa, requested my friend, Mr. Lawrence M. Lambe, artist to the Survey, who was well acquainted with that locality, to recommend me to a good hotel. He at once made a pencil sketch of a little, stout, dumpy man, and said: "That is something like my friend, Mr. C., who keeps the best hotel at Murray Bay."

Monday morning, June 27th, I boarded the fine S. S. Saguenay, and after a delightful trip arrived at this charming salt-water summer resort at 4 p. m. With Mr. Lambe's sketch in my hand I looked over the gay crowd of summer visitors, who always flock to the wharf to see new arrivals, and sure enough among this group was our little dumpy man. The sketch was so like him that when I presented it to Mr. C. and said, that is the man I am looking for, he fairly shook with laughter, said, "that's me," and requested to be allowed to keep the sketch.

REMINISCENCES

The days were beautiful, and in the evenings when music and laughter prevailed, one thought of Longfellow's words :—

> " And the nights shall be filled with music,
> And the cares that infest the day,
> Shall fold their tents like the Arabs,
> And as silently steal away."

Here, at Murray Bay, as at most of the salt water summer resorts below Quebec, the man with the hammer finds the rocks more interesting than the festivities of a gay hotel, and wanders forth into the valley of Murray Bay River to search for new fossil evidence, which there is always a chance of obtaining in any of nature's rock storehouses.

But to one interested in the Laurentian formation of the Lower St. Lawrence River, a tramp over the sea-splashed rocks between Murray Bay and Bay St. Paul, gives much pleasure, for here, if one is fitting up a cabinet of Laurentian rocks, he can find a charming variety. But while at this work, he must not forget—like the writer of these notes did—and get cut off by the tide, and have to remain in an awkward position till ebb tide.

Before leaving Quebec for Ottawa, I was requested to again visit officially the Joggins Coast, N. S. In previous pages of these memoirs I have spoken of the magnificent section of carboniferous rocks which occupy the south coast of

AMONG THE ROCKS.

Cumberland Basin, or the head of the Bay of Fundy.

The writings of Sir William Logan, Sir William Dawson. and other celebrated geologists, have told us of the remarkable fauna and flora of these coast rocks, and given us sections and descriptions of the Joggins Coast deposits till one would almost fancy the geological structure, and palæontological evidence was exhausted, but that is not so, for during this my last visit to this ever-interesting locality it was my good fortune to discover two very large unio-like shells. They are fresh water bivalves, one measures a little less than eight inches in length, and the other a little more than eight inches. They were found in the rocks belonging to the productive coal measures, with *Sigillaria*, *Lepidodendron* and other Carboniferous fossil plants, and an almost perfect lower jaw of a reptile whose largest tooth was half an inch long. It was probably the maxillary bone *Baphetes planiceps*. Owen, (Acadian Geology, Dawson, p. 360). Like many fossil bones of the carboniferous formation this was so fractured that it could not be preserved with the means then at hand.

One of these remarkable fossil bivalves has been described and figured by Mr. J. F. Whiteaves, F. R. S. C., F. G. S., palæontologist to the Geological Survey of Canada, in Transactions of the Royal Society of Canada. It received the name *Asthedonta Westoni*, being dedicated to the discoverer.

My friend, Mr. G. F. Matthew of St. John, N. B., who has done much good geological work for the Survey, having heard of this Carboniferous fossil bivalve, writes: " Let me congratulate you on the giant clam you have found; quite a respectable companion to him of the Pacific ocean and a head and shoulders taller than any modern fresh water beast of our country."

Another important fossil find was made this season in the productive coal measures of the Joggins Coast. Sunday July 15th while strolling along the shore, in the vicinity of Hardscrabble Wharf, with my coal miner friend, Mr. James Devine—who has a sharp eye for fossils—we discovered on the surface of a large block of sandstone which had fallen from one of the upper beds of the cliff, the tracks of a small reptile. It is probable that these are the footprints of the small saurian reptile, or lizard *Hylonomus Lyelli*, Dawson, (Acadian Geology, Dawson 1863 p. 373.) It is probable that the rami of the small reptile which made these tracks did not exceed half an inch in length. These, almost the first vertebrate animals which appeared on this earth, were the small reptiles whose bones Sir William Dawson has found in the base of erect fossil trees belonging to the genus *Sigillaria*. These delicate little lizards, as

AMONG THE ROCKS.

the name *Hylonomus* indicates, were forest dwellers, and lived chiefly in the base of the coneiferous trees which formed a large portion of the forests of coal period. The slab of sandstone just spoken of—now in the Geological Survey museum—containing the footprints, was obtained with much difficulty, apart from our having to carry it for more than a mile over a rough sea coast, but one is always ready to risk a little to obtain a good geological specimen.

The study of the Carboniferous system in Cumberland county, Nova Scotia, is very fascinating; the great variety of rocks including sandstones shales, conglomerates, limestones, clays, gravels, etc., with their vastly important bituminous coal beds, and great cliffs of sandstone peculiarly adapted for grindstones, thousands of which are yearly manufactured close to the sea shore, a few miles distant from the Joggins coal mines. These rocks which in Canada are spread over a good portion of New Brunswick and Nova Scotia have an estimated thickness of about 11,000 feet, not including the Permian beds. They were formed in the ancient carboniferous seas, and contain, besides a very large number of species of fossil plants, the bones of extinct fish and reptiles; the foot prints of great Saurians and, as we have just seen, the tracks of small lizards whose whole body would not measure more than the length of a claw of the great Saurian who has, so far as we know, only left his footprints behind.

From the ever-interesting shores of the Joggins coast, where the sea lashes the great sandstone cliffs, which as I have already said occasionally exhibit sections of erect fossil trees, I paid my initial visit to Spring Hill. Arriving there July 24th, I at once visited the coal mines and through the courtesy of Mr. Alex. McInnes got a slight idea of the extensive mining then going on.

The village or town of Spring Hill is situated inland a few miles from the Joggins coast, reached now by railway. Geologically, above ground, this is not an interesting locality; I saw no rock exposures near the village, and the only chance of examining the geological strata is in the mines, or on the dumping ground where one can get some idea of the underlying deposits, and occasionally pick up a good fossil plant or perhaps a section of a "petrified" trunk of a tree such as those found in the cliffs of the Joggins coast.

I had only been at Spring Hill a few hours, and was getting much interested in the deposits on the dumping ground where several good fossil ferns were found, when suddenly a severe sickness overtook me, resulting probably from some fish I had eaten for dinner. Attended by Dr. Cove and the cook of the hotel I remained in bed five days—the longest five days I ever spent. But there was a silver lining to my dark cloud, which was represented by wonderful extempore music created by the host of the hotel, a young man who had recently become proprietor of this famous old hostelry.

To Mr. Alex. McInnes, manager of the Spring Hill coal mines, and Dr. Cove, M.A., my thanks were specially due for many kindesses, especially for collecting a number of fossil plants for me during my short but almost fatal sickness which lasted some days after I again returned to Ottawa. It was this sickness and general debility that caused me to apply for superannuation, which after a further leave of absence for four months, during which my time, whenever able, was spent on the Quebec rocks in the interest of the Survey—was granted on the 1st of July 1894.

A few more words before closing these rambling remarks. During my 35 years' connection with the Geological Survey of Canada changes have constantly taken place, and constant improvements in this Institution have been made, especially since the Dominion Government decided to remove the Survey to Ottawa. The palæontological branch has (like all the other branches) prospered under the able chief of that department Mr. J. F. Whiteaves and his assistants, one of whom it has been my privilege to be since the death of the former palæontologist, Mr. E. Billings. It has not been as some members of the Survey prophesied—"the beginning of the end,"

REMINISCENCES

but rather the beginning of a new epoch in the history of this important institution, which I feel sure will continue to prosper under the able management of its present director, Dr. G. M. Dawson. I may mention here that while the Survey was in Montreal the largest number of visitors, to the museum, in any year, was 1652. In 1897 (this year) the number has increased to 32,357. Year by year these numbers have risen and will probably be doubled when a proper building is provided for displaying the thousands of specimens now stored away.

Passing through the various departments of the Survey one is constantly reminded of absent friends who have long passed away, but whose faithful works in the interests of geology will always remain as monuments of their industry and love of their chosen pursuits.

In previous pages of these notes I have spoken of some of these diligent workers in our Canadian Geological field, and now to the present members of the Geological staff: Dr. G. M. Dawson, Dr. Bell, Mr. J. F. Whiteaves, Dr. Ami, Mr. Fletcher, Mr. Lambe, Mr. Dowling, Prof. Macoun, and others of the Survey, who are diligently continuing the good work of their predecessors, I tender my sincere thanks for friendship and valued counsel.

www.ingramcontent.com/pod-product-compliance
Lightning Source LLC
Chambersburg PA
CBHW020233240426
43672CB00006B/516